D0820403

Applied Mathematical Sciences
Volume 153

Editors
S.S. Antman   J.E. Marsden   L. Sirovich

Advisors
J.K. Hale   P. Holmes   J. Keener
J. Keller   B.J. Matkowksy   A. Mielke
C.S. Peskin   K.R. Sreenivasan

Springer

Applied Mathematical Sciences

(continued following index)

Stanley Osher   Ronald Fedkiw

Level Set Methods and Dynamic Implicit Surfaces

 Springer

Stanley Osher
Department of Mathematics
University of California at Los Angeles
Los Angeles, CA 90095-1555
USA
sjo@math.ucla.edu

Ronald Fedkiw
Department of Computer Science
Stanford University
Stanford, CA 94305-9020
USA
fedkiw@cs.stanford.edu

Editors:

S.S. Antman
Department of Mathematics
and
Institute for Physical Science
 and Technology
University of Maryland
College Park, MD 20742-4015
USA
ssa@math.umd.edu

J.E. Marsden
Control and Dynamical
 Systems, 107-81
California Institute of
 Technology
Pasadena, CA 91125
USA
marsden@cds.caltech.edu

L. Sirovich
Division of Applied
 Mathematics
Brown University
Providence, RI 02912
USA
chico@camelot.mssm.edu

Cover photos: Top left and right, hand and rat brain—Duc Nguyen and Hong-Kai Zhao. Center campfire—Duc Nguyen and Nick Rasmussen and Industrial Light and Magic. Lower left and center, water glasses—Steve Marschner and Dough Enright.

Mathematics Subject Classification (2000): 65Mxx, 65C20, 65D17, 65-02, 65V10, 73V

Library of Congress Cataloging-in-Publication Data
Osher, Stanley.
    Level set methods and dynamic implicit surfaces / Stanley Osher, Ronald Fedkiw
        p. cm. – (Applied mathematical sciences; 153)
    Includes bibliographical references and index.
    ISBN 0-387-95482-1 (alk. paper)
    1. Level set methods. 2. Implicit functions. I. Fedkiw, Ronald P., 1968– II. Title.
    III. Applied mathematical sciences (Springer-Verlag New York Inc.); v. 153
    QA1.A647 vol. 153
    [QC173.4]
    510s–dc21
    [515′. 8]

ISBN 978-0-387-95482-0   2002020939

Printed on acid-free paper.

© 2003 Springer Science+Business Media, LLC
All rights reserved. This work may not be translated or copied in whole or in part without the written permission of the publisher (Springer Science+Business Media, LLC, 233 Spring Street, New York, NY 10013, USA), except for brief excerpts in connection with reviews or scholarly analysis. Use in connection with any form of information storage and retrieval, electronic adaptation, computer software, or by similar or dissimilar methodology now known or hereafter developed is forbidden. The use in this publication of trade names, trademarks, service marks, and similar terms, even if they are not identified as such, is not to be taken as an expression of opinion as to whether or not they are subject to proprietary rights.

Printed in the United States of America.

9 8 7 6 5 4

springer.com

Dedicated with love to Katy, Brittany, and Bobbie

Preface

Scope, Aims, and Audiences

This book, *Level Set Methods and Dynamic Implicit Surfaces* is designed to serve two purposes:

Parts I and II introduce the reader to implicit surfaces and level set methods. We have used these chapters to teach introductory courses on the material to students with little more than a fundamental math background. No prior knowledge of partial differential equations or numerical analysis is required. These first eight chapters include enough detailed information to allow students to create working level set codes from scratch.

Parts III and IV of this book are based on a series of papers published by us and our colleagues. For the sake of brevity, a few details have been occasionally omitted. These chapters do include thorough explanations and enough of the significant details along with the appropriate references to allow the reader to get a firm grasp on the material.

This book is an introduction to the subject. We have given examples of the utility of the method to a diverse (but by no means complete) collection of application areas. We have also tried to give complete numerical recipes and a self-contained course in the appropriate numerical analysis. We believe that this book will enable users to apply the techniques presented here to real problems.

The level set method has been used in a rapidly growing number of areas, far too many to be represented here. These include epitaxial growth, optimal design, CAD, MEMS, optimal control, and others where the simulation of moving interfaces plays a key role in the problem to be solved. A search of "level set methods" on the Google website (which gave over 2,700 responses as of May 2002) will give an interested reader some idea of the scope and utility of the method. In addition, some exciting advances in the technology have been made since we began writing this book. We hope to cover many of these topics in a future edition. In the meantime you can find some exciting animations and moving images as well as links to more relevant research papers via our personal web sites: `http://graphics.stanford.edu/~fedkiw` and `http://www.math.ucla.edu/~sjo/`.

Acknowledgments

Many people have helped us in this effort. We thank the following colleagues in particular: Steve Marschner, Paul Romburg, Gary Hewer, and Steve Ruuth for proofreading parts of the manuscript, Peter Smereka and Li-Tien Cheng for providing figures for the chapter on Codimension-Two Objects, Myungjoo Kang for providing figures for the chapter on Motion Involving Mean Curvature and Motion in the Normal Direction, Antonio Marquina and Frederic Gibou for help with the chapter on Image Restoration, Hong-Kai Zhao for help with chapter 13, Reconstruction of Surfaces from Unorganized Data Points, and Luminita Vese for help with the chapter on Snakes, Active Contours, and Segmentation. We particularly thank Barry Merriman for his extremely valuable collaboration on much of the research described here. Of course we have benefitted immensely from collaborations and discussions with far too many people to mention. We hope these colleagues and friends forgive us for omitting their names.

We would like to thank the following agencies for their support during this period: ONR, AFOSR, NSF, ARO, and DARPA. We are particularly grateful to Dr. Wen Masters of ONR for suggesting and believing in this project and for all of her encouragement during some of the more difficult times.

Finally, we thank our families and friends for putting up with us during this exciting, but stressful period.

Los Angeles, California   Stanley Osher
Stanford, California   Ronald Fedkiw

Contents

Part I
Implicit Surfaces

In the next two chapters we introduce implicit surfaces and illustrate a number of useful properties, focusing on those that will be of use to us later in the text. A good general review can be found in [16]. In the first chapter we discuss those properties that are true for a general implicit representation. In the second chapter we introduce the notion of a signed distance function with a Euclidean distance metric and a "±" sign used to indicate the inside and outside of the surface.

1
Implicit Functions

1.1 Points

In one spatial dimension, suppose we divide the real line into three distinct pieces using the points $x = -1$ and $x = 1$. That is, we define $(-\infty, -1)$, $(-1, 1)$, and $(1, \infty)$ as three separate subdomains of interest, although we regard the first and third as two disjoint pieces of the same region. We refer to $\Omega^- = (-1, 1)$ as the *inside* portion of the domain and $\Omega^+ = (-\infty, -1) \cup (1, \infty)$ as the *outside* portion of the domain. The border between the inside and the outside consists of the two points $\partial\Omega = \{-1, 1\}$ and is called the *interface*. In one spatial dimension, the inside and outside regions are one-dimensional objects, while the interface is less than one-dimensional. In fact, the points making up the interface are zero-dimensional. More generally, in $\Re^n$, subdomains are n-dimensional, while the interface has dimension $n - 1$. We say that the interface has codimension one.

In an *explicit* interface representation one explicitly writes down the points that belong to the interface as we did above when defining $\partial\Omega = \{-1, 1\}$. Alternatively, an *implicit* interface representation defines the interface as the isocontour of some function. For example, the zero isocontour of $\phi(x) = x^2 - 1$ is the set of all points where $\phi(x) = 0$; i.e., it is exactly $\partial\Omega = \{-1, 1\}$. This is shown in Figure 1.1. Note that the implicit function $\phi(x)$ is defined throughout the one-dimensional domain, while the isocontour defining the interface is one dimension lower. More generally, in $\Re^n$, the implicit function $\phi(\vec{x})$ is defined on all $\vec{x} \in \Re^n$, and its isocontour has dimension $n - 1$. Initially, the implicit representation might seem

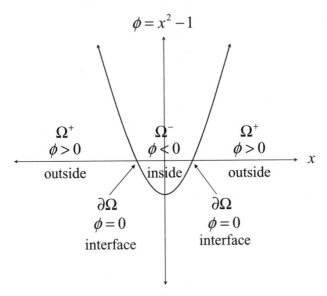

Figure 1.1. Implicit function $\phi(x) = x^2 - 1$ defining the regions Ω^- and Ω^+ as well as the boundary $\partial\Omega$

wasteful, since the implicit function $\phi(\vec{x})$ is defined on all of $\Re^n$, while the interface has only dimension $n - 1$. However, we will see that a number of very powerful tools are readily available when we use this representation.

Above, we chose the $\phi(x) = 0$ isocontour to represent the lower-dimensional interface, but there is nothing special about the zero isocontour. For example, the $\hat{\phi}(x) = 1$ isocontour of $\hat{\phi}(x) = x^2$, defines the same interface, $\partial\Omega = \{-1, 1\}$. In general, for any function $\hat{\phi}(\vec{x})$ and an arbitrary isocontour $\hat{\phi}(\vec{x}) = a$ for some scalar $a \in \Re$, we can define $\phi(\vec{x}) = \hat{\phi}(\vec{x}) - a$, so that the $\phi(\vec{x}) = 0$ isocontour of ϕ is identical to the $\hat{\phi}(\vec{x}) = a$ isocontour of $\hat{\phi}$. In addition, the functions ϕ and $\hat{\phi}$ have identical properties up to a scalar translation a. Moreover, the partial derivatives of ϕ are the same as the partial derivatives of $\hat{\phi}$, since the scalar vanishes upon differentiation. Thus, throughout the text all of our implicit functions $\phi(\vec{x})$ will be defined so that the $\phi(\vec{x}) = 0$ isocontour represents the interface (unless otherwise specified).

1.2   Curves

In two spatial dimensions, our lower-dimensional interface is a curve that separates $\Re^2$ into separate subdomains with nonzero areas. Here we are limiting our interface curves to those that are *closed*, so that they have clearly defined interior and exterior regions. As an example, consider $\phi(\vec{x}) =$

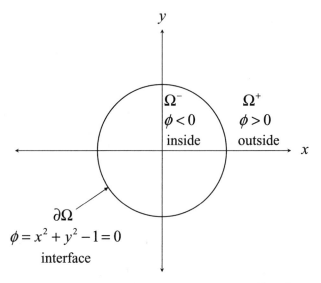

Figure 1.2. Implicit representation of the curve $x^2 + y^2 = 1$.

$x^2 + y^2 - 1$, where the interface defined by the $\phi(\vec{x}) = 0$ isocontour is the unit circle defined by $\partial\Omega = \{\vec{x} \mid |\vec{x}| = 1\}$. The interior region is the unit open disk $\Omega^- = \{\vec{x} \mid |\vec{x}| < 1\}$, and the exterior region is $\Omega^+ = \{\vec{x} \mid |\vec{x}| > 1\}$. These regions are depicted in Figure 1.2. The explicit representation of this interface is simply the unit circle defined by $\partial\Omega = \{\vec{x} \mid |\vec{x}| = 1\}$.

In two spatial dimensions, the explicit interface definition needs to specify all the points on a curve. While in this case it is easy to do, it can be somewhat more difficult for general curves. In general, one needs to parameterize the curve with a vector function $\vec{x}(s)$, where the parameter s is in $[s_o, s_f]$. The condition that the curve be closed implies that $\vec{x}(s_o) = \vec{x}(s_f)$.

While it is convenient to use analytical descriptions as we have done so far, complicated two-dimensional curves do not generally have such simple representations. A convenient way of approximating an explicit representation is to discretize the parameter s into a finite set of points $s_o < \cdots < s_{i-1} < s_i < s_{i+1} < \cdots < s_f$, where the subintervals $[s_i, s_{i+1}]$ are not necessarily of equal size. For each point s_i in parameter space, we then store the corresponding two-dimensional location of the curve denoted by $\vec{x}(s_i)$. As the number of points in the discretized parameter space is increased, so is the resolution (detail) of the two-dimensional curve.

The implicit representation can be stored with a discretization as well, except now one needs to discretize all of $\Re^2$, which is impractical, since it is unbounded. Instead, we discretize a bounded subdomain $D \subset \Re^2$. Within this domain, we choose a finite set of points (x_i, y_i) for $i = 1, \ldots, N$ to discretely approximate the implicit function ϕ. This illustrates a drawback of the implicit surface representation. Instead of resolving a one-dimensional interval $[s_o, s_f]$, one needs to resolve a two-dimensional region D. More generally, in $\Re^n$, a discretization of an explicit representation needs to resolve only an $(n-1)$-dimensional set, while a discretization of an implicit representation needs to resolve an n-dimensional set. This can be avoided, in part, by placing all the points $\vec{x}$ very close to the interface, leaving the rest of D unresolved. Since only the $\phi(\vec{x}) = 0$ isocontour is important, only the points $\vec{x}$ near this isocontour are actually needed to accurately represent the interface. The rest of D is unimportant. Clustering points near the interface is a *local* approach to discretizing implicit representations. (We will give more details about local approaches later.) Once we have chosen the set of points that make up our discretization, we store the values of the implicit function $\phi(\vec{x})$ at each of these points.

Neither the explicit nor the implicit discretization tells us where the interface is located. Instead, they both give information at sample locations. In the explicit representation, we know the location of a finite set of points on the curve, but do not know the location of the remaining infinite set of points (on the curve). Usually, interpolation is used to approximate the location of points not represented in the discretization. For example, piecewise polynomial interpolation can be used to determine the shape of the interface between the data points. Splines are usually appropriate for this. Similarly, in the implicit representation we know the values of the implicit function ϕ at only a finite number of points and need to use interpolation to find the values of ϕ elsewhere. Even worse, here we may not know the location of any of the points on the interface, unless we have luckily chosen data points $\vec{x}$ where $\phi(\vec{x})$ is exactly equal to zero. In order to locate the interface, the $\phi(\vec{x}) = 0$ isocontour needs to be interpolated from the known values of ϕ at the data points. This is a rather standard procedure accomplished by a variety of contour plotting routines.

The set of data points where the implicit function ϕ is defined is called a *grid*. There are many ways of choosing the points in a grid, and these lead to a number of different types of grids, e.g., unstructured, adaptive, curvilinear. By far, the most popular grids, are Cartesian grids defined as $\{(x_i, y_j) \mid 1 \leq i \leq m, 1 \leq j \leq n\}$. The natural orderings of the x_i and y_j are usually used for convenience. That is, $x_1 < \cdots < x_{i-1} < x_i < x_{i+1} < \cdots < x_m$ and $y_1 < \cdots < y_{j-1} < y_j < y_{j+1} < \cdots < y_n$. In a *uniform* Cartesian grid, all the subintervals $[x_i, x_{i+1}]$ are equal in size, and we set $\triangle x = x_{i+1} - x_i$. Likewise, all the subintervals $[y_j, y_{j+1}]$ are equal in size, and we set $\triangle y = y_{j+1} - y_j$. Furthermore, it is usually convenient to choose $\triangle x = \triangle y$ so that the approximation errors are the same in the x-direction as they are in the y-direction. By definition, Cartesian grids imply the use of a rectangular domain $D = [x_1, x_m] \times [y_1, y_n]$. Again, since ϕ is important only near the interface, a local approach would indicate that many of the grid points are not needed, and the implicit representation can be optimized by storing only a subset of a uniform Cartesian grid. The Cartesian grid points that are not sufficiently near the interface can be discarded.

We pause for a moment to consider the discretization of the one-dimensional problem. There, since the explicit representation is merely a set of points, it is trivial to record the exact interface position, and no discretization or parameterization is needed. However, the implicit representation must be discretized if ϕ is not a known analytic function. A typical discretization consists of a set of points $x_1 < \cdots < x_{i-1} < x_i < x_{i+1} < \cdots < x_m$ on a subdomain $D = [x_1, x_m]$ of $\Re$. Again, it is usually useful to use a uniform grid, and only the grid points near the interface need to be stored.

1.3 Surfaces

In three spatial dimensions the lower-dimensional interface is a surface that separates $\Re^3$ into separate subdomains with nonzero volumes. Again, we consider only closed surfaces with clearly defined interior and exterior regions. As an example, consider $\phi(\vec{x}) = x^2 + y^2 + z^2 - 1$, where the interface is defined by the $\phi(\vec{x}) = 0$ isocontour, which is the boundary of the unit sphere defined as $\partial\Omega = \{\vec{x} \mid |\vec{x}| = 1\}$. The interior region is the open unit sphere $\Omega^- = \{\vec{x} \mid |\vec{x}| < 1\}$, and the exterior region is $\Omega^+ = \{\vec{x} \mid |\vec{x}| > 1\}$. The explicit representation of the interface is $\partial\Omega = \{\vec{x} \mid |\vec{x}| = 1\}$.

For complicated surfaces with no analytic representation, we again need to use a discretization. In three spatial dimensions the explicit representation can be quite difficult to discretize. One needs to choose a number of points on the two-dimensional surface and record their connectivity. In two spatial dimensions, connectivity was determined based on the ordering, i.e., $\vec{x}(s_i)$ is connected to $\vec{x}(s_{i-1})$ and $\vec{x}(s_{i+1})$. In three spatial dimensions connectivity is less straightforward. If the exact surface and its connectivity are known, it is simple to tile the surface with triangles whose vertices lie on the interface and whose edges indicate connectivity. On the other hand, if connectivity is not known, it can be quite difficult to determine, and even some of the most popular algorithms can produce surprisingly inaccurate surface representations, e.g., surfaces with holes.

Connectivity can change for *dynamic* implicit surfaces, i.e., surfaces that are moving around. As an example, consider the splashing water surface in a swimming pool full of children. Here, connectivity is not a "one-time" issue dealt with in constructing an explicit representation of the surface. Instead, it must be resolved over and over again every time pieces of the surface *merge* together or *pinch* apart. In two spatial dimensions the task is more manageable, since merging can be accomplished by taking two one-dimensional parameterizations, s_i and $\hat{s}_i$, and combining them into a single one-dimensional parameterization. Pinching apart is accomplished by splitting a single one-dimensional parameterization into two separate one-dimensional parameterizations. In three spatial dimensions the "interface surgery" needed for merging and pinching is much more complex, leading to a number of difficulties including, for example, holes in the surface.

One of the nicest properties of implicit surfaces is that connectivity does not need to be determined for the discretization. A uniform Cartesian grid $\{(x_i, y_j, z_k) \mid 1 \leq i \leq m, 1 \leq j \leq n, 1 \leq k \leq p\}$ can be used along with straightforward generalizations of the technology from two spatial dimensions. Possibly the most powerful aspect of implicit surfaces is that it is straightforward to go from two spatial dimensions to three spatial dimensions (or even more).

1.4   Geometry Toolbox

Implicit interface representations include some very powerful geometric tools. For example, since we have designated the $\phi(\vec{x}) = 0$ isocontour as the interface, we can determine which side of the interface a point is on simply by looking at the local sign of ϕ. That is, $\vec{x}_o$ is inside the interface when $\phi(\vec{x}_o) < 0$, outside the interface when $\phi(\vec{x}_o) > 0$, and on the interface when $\phi(\vec{x}_o) = 0$. With an explicit representation of the interface it can be difficult to determine whether a point is inside or outside the interface. A standard procedure for doing this is to cast a ray from the point in question to some far-off place that is known to be outside the interface. Then if the ray intersects the interface an even number of times, the point is outside the interface. Otherwise, the ray intersects the interface an odd number of times, and the point is inside the interface. Obviously, it is more convenient simply to evaluate ϕ at the point $\vec{x}_o$. In the discrete case, i.e., when the implicit function is given by its values at a finite number of data points, interpolation can be used to estimate $\phi(\vec{x}_o)$ using the values of ϕ at the known sample points. For example, on our Cartesian grid, linear, bilinear, and trilinear interpolation can be used in one, two, and three spatial dimensions, respectively.

Numerical interpolation produces errors in the estimate of ϕ. This can lead to erroneously designating inside points as outside points and vice versa. At first glance these errors might seem disastrous, but in reality they amount to perturbing (or moving) the interface away from its exact position. If these interface perturbations are small, their effects may be minor, and a perturbed interface might be acceptable. In fact, most numerical methods depend on the fact that the results are stable in the presence of small perturbations. If this is not true, then the problem under consideration is probably ill-posed, and numerical methods should be used only with extreme caution (and suspicion). These interface perturbation errors decrease as the number of sample points increases, implying that the exact answer could hypothetically be computed as the number of sample points is increased to infinity. Again, this is the basis for most numerical methods.

While one cannot increase the number of grid points to infinity, desirable solutions can be obtained for many problems with a practical number of grid points. Throughout the text we will make a number of numerical approximations with errors proportional to the size of a Cartesian mesh cell, i.e., $\triangle x$ (or $(\triangle x)^r$). If the implicit function is smooth enough and well resolved by the grid, these estimates will be appropriate. Otherwise, these errors might be rather large. Obviously, this means that we would like our implicit function to be as smooth as possible. In the next chapter we discuss using a *signed distance function* to represent the surface. This turns out to be a good choice, since steep and flat gradients as well as rapidly changing features are avoided as much as possible.

Implicit functions make both simple Boolean operations and more advanced constructive solid geometry (CSG) operations easy to apply. This is important, for example, in computer-aided design (CAD). If ϕ_1 and ϕ_2 are two different implicit functions, then $\phi(\vec{x}) = \min(\phi_1(\vec{x}), \phi_2(\vec{x}))$ is the implicit function representing the *union* of the interior regions of ϕ_1 and ϕ_2. Similarly, $\phi(\vec{x}) = \max(\phi_1(\vec{x}), \phi_2(\vec{x}))$ is the implicit function representing the *intersection* of the interior regions of ϕ_1 and ϕ_2. The *complement* of $\phi_1(\vec{x})$ can be defined by $\phi(\vec{x}) = -\phi_1(\vec{x})$. Also, $\phi(\vec{x}) = \max(\phi_1(\vec{x}), -\phi_2(\vec{x}))$ represents the region obtained by subtracting the interior of ϕ_2 from the interior of ϕ_1.

The *gradient* of the implicit function is defined as

$$\nabla\phi = \left(\frac{\partial\phi}{\partial x}, \frac{\partial\phi}{\partial y}, \frac{\partial\phi}{\partial z}\right). \tag{1.1}$$

The gradient $\nabla\phi$ is perpendicular to the isocontours of ϕ and points in the direction of increasing ϕ. Therefore, if $\vec{x}_o$ is a point on the zero isocontour of ϕ, i.e., a point on the interface, then $\nabla\phi$ evaluated at $\vec{x}_o$ is a vector that points in the same direction as the local unit (outward) normal $\vec{N}$ to the interface. Thus, the unit (outward) *normal* is

$$\vec{N} = \frac{\nabla\phi}{|\nabla\phi|} \tag{1.2}$$

for points on the interface.

Since the implicit representation of the interface embeds the interface in a domain of one higher-dimension, it will be useful to have as much information as possible representable on the higher-dimensional domain. For example, instead of defining the unit normal $\vec{N}$ by equation (1.2) for points on the interface only, we use equation (1.2) to define a function $\vec{N}$ everywhere on the domain. This embeds the normal in a function $\vec{N}$ defined on the entire domain that agrees with the normal for points on the interface. Figure 1.3 shows a few isocontours of our two-dimensional example $\phi(\vec{x}) = x^2 + y^2 - 1$ along with some representative normals.

Consider the one-dimensional example $\phi(x) = x^2 - 1$, where $\vec{N}$ is defined by equation (1.2) as $\vec{N} = x/|x|$. Here, $\vec{N}$ points to the right for all $x > 0$

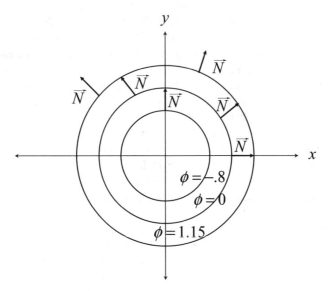

Figure 1.3. A few isocontours of our two-dimensional example $\phi(\vec{x}) = x^2 + y^2 - 1$ along with some representative normals.

including $x = 1$, where the interface normal is $\vec{N} = 1$, and $\vec{N}$ points to the left for all $x < 0$ including $x = -1$, where the interface normal is $\vec{N} = -1$. The normal is undefined at $x = 0$, since the denominator of equation (1.2) vanishes. This can be problematic in general, but can be avoided with a number of techniques. For example, at $x = 0$ we could simply define $\vec{N}$ as either $\vec{N} = 1$ or $\vec{N} = -1$. Our two- and three-dimensional examples (above) show similar degenerate behavior at $\vec{x} = \vec{0}$, where all partial derivatives vanish. Again, a simple technique for evaluating (1.2) at these points is just to pick an arbitrary direction for the normal. Note that the standard trick of adding a small $\epsilon > 0$ to the denominator of equation (1.2) can be a bad idea in general, since it produces a normal with $|\vec{N}| \neq 1$. In fact, when the denominator in equation (1.2) is zero, so is the numerator, making $\vec{N} = \vec{0}$ when a small $\epsilon > 0$ is used in the denominator. (While setting $\vec{N} = \vec{0}$ is not always disastrous, caution is advised.)

On our Cartesian grid, the derivatives in equation (1.2) need to be approximated, for example using finite difference techniques. We can use a first-order accurate forward difference

$$\frac{\partial \phi}{\partial x} \approx \frac{\phi_{i+1} - \phi_i}{\triangle x} \tag{1.3}$$

abbreviated as $D^+\phi$, a first-order accurate backward difference

$$\frac{\partial \phi}{\partial x} \approx \frac{\phi_i - \phi_{i-1}}{\triangle x} \tag{1.4}$$

abbreviated as $D^-\phi$, or a second-order accurate central difference

$$\frac{\partial \phi}{\partial x} \approx \frac{\phi_{i+1} - \phi_{i-1}}{2\triangle x} \tag{1.5}$$

abbreviated as $D^o\phi$. (The j and k indices have been suppressed in the above formulas.) The formulas for the derivatives in the y and z directions are obtained through symmetry. These simple formulas are by no means exhaustive, and we will discuss more ways of approximating derivatives later in the text.

When all numerically calculated finite differences are identically zero, the denominator of equation (1.2) vanishes. As in the analytic case, we can simply randomly choose a normal. Here, however, randomly choosing a normal is somewhat justified, since it is equivalent to randomly perturbing the values of ϕ on our Cartesian mesh by values near round-off error. These small changes in the values of ϕ are dominated by the local approximation errors in the finite difference formula for the derivatives. Consider a discretized version of our one-dimensional example $\phi(x) = x^2 - 1$, and suppose that grid points exist at $x_{i-1} = -\triangle x$, $x_i = 0$, and $x_{i+1} = \triangle x$ with exact values of ϕ defined as $\phi_{i-1} = \triangle x^2 - 1$, $\phi_i = -1$, and $\phi_{i+1} = \triangle x^2 - 1$, respectively. The forward difference formula gives $\vec{N}_i = 1$, the backward difference formula gives $\vec{N}_i = -1$, and the central difference formula cannot be used, since $D^o\phi = 0$ at $x_i = 0$. However, simply perturbing ϕ_{i+1} to $\triangle x^2 - 1 + \epsilon$ for any small $\epsilon > 0$ (even round-off error) gives $D^o\phi \neq 0$ and $\vec{N}_i = 1$. Similarly, perturbing ϕ_{i-1} to $\triangle x^2 - 1 + \epsilon$ gives $\vec{N}_i = -1$. Thus, for any approach that is stable under small perturbations of the data, it is acceptable to randomly choose $\vec{N}$ when the denominator of equation (1.2) vanishes. Similarly in our two- and three-dimensional examples, $\vec{N} = \vec{x}/|\vec{x}|$ everywhere except at $\vec{x} = \vec{0}$, where equation (1.2) is not defined and we are free to choose it arbitrarily. The arbitrary normal at the origin in the one-dimensional case lines up with the normals to either the right, where $\vec{N} = 1$, or to the left, where $\vec{N} = -1$. Similarly, in two and three spatial dimensions, an arbitrarily chosen normal at $\vec{x} = \vec{0}$ lines up with other nearby normals. This is always the case, since the normals near the origin point outward in every possible direction.

If ϕ is a smooth well-behaved function, then an approximation to the value of the normal at the interface can be obtained from the values of $\vec{N}$ computed at the nodes of our Cartesian mesh. That is, given a point $\vec{x}_o$ on the interface, one can estimate the unit outward normal at $\vec{x}_o$ by interpolating the values of $\vec{N}$ from the Cartesian mesh to the point $\vec{x}_o$. If one is using forward, backward, or central differences, then linear (bilinear or trilinear) interpolation is usually good enough. However, higher-order accurate formulas can be used if desired. This interpolation procedure requires that ϕ be well behaved, implying that we should be careful in how we choose ϕ. For example, it would be unwise to choose an implicit function ϕ with unnecessary oscillations or steep (or flat) gradients. Again, a good choice for ϕ turns out to be the signed distance function discussed in the next chapter.

The mean *curvature* of the interface is defined as the divergence of the normal $\vec{N} = (n_1, n_2, n_3)$,

$$\kappa = \nabla \cdot \vec{N} = \frac{\partial n_1}{\partial x} + \frac{\partial n_2}{\partial y} + \frac{\partial n_3}{\partial z}, \tag{1.6}$$

so that $\kappa > 0$ for *convex* regions, $\kappa < 0$ for *concave* regions, and $\kappa = 0$ for a plane; see Figure 1.4. While one could simply use finite differences to compute the derivatives of the components of the normal in equation (1.6), it is usually more convenient, compact, and accurate to calculate the curvature directly from the values of ϕ. Substituting equation (1.2) into equation (1.6) gives

$$\kappa = \nabla \cdot \left(\frac{\nabla \phi}{|\nabla \phi|} \right), \tag{1.7}$$

so that we can write the curvature as

$$\kappa = \left(\phi_x^2 \phi_{yy} - 2\phi_x \phi_y \phi_{xy} + \phi_y^2 \phi_{xx} + \phi_x^2 \phi_{zz} - 2\phi_x \phi_z \phi_{xz} + \phi_z^2 \phi_{xx} \right.$$
$$\left. + \phi_y^2 \phi_{zz} - 2\phi_y \phi_z \phi_{yz} + \phi_z^2 \phi_{yy} \right) / |\nabla \phi|^3 \tag{1.8}$$

in terms of the first and second derivatives of ϕ. A second-order accurate finite difference formula for ϕ_{xx}, the second partial derivative of ϕ in the x direction, is given by

$$\frac{\partial^2 \phi}{\partial x^2} \approx \frac{\phi_{i+1} - 2\phi_i + \phi_{i-1}}{\Delta x^2}, \tag{1.9}$$

abbreviated as $D_x^+ D_x^- \phi$, or equivalently, $D_x^- D_x^+ \phi$. Here D^+ and D^- are defined as in equations (1.3) and (1.4), respectively, and the x subscript indicates that the finite difference is evaluated in the x direction. A second-order accurate finite difference formula for ϕ_{xy} is given by $D_x^o D_y^o \phi$, or equivalently, $D_y^o D_x^o \phi$. The other second derivatives in equation (1.8) are defined in a manner similar to either ϕ_{xx} or ϕ_{xy}.

In our one-dimensional example, $\phi(x) = x^2 - 1$, $\kappa = 0$ everywhere except at the origin, where equation (1.7) is undefined. Thus, the origin, is a removable singularity, and we can define $\kappa = 0$ everywhere. Interfaces in one spatial dimension are models of planes in three dimensions (assuming that the unmodeled directions have uniform data). Therefore, using $\kappa = 0$ everywhere is a consistent model. In our two- and three-dimensional examples above, $\kappa = \frac{1}{|\vec{x}|}$ and $\kappa = \frac{2}{|\vec{x}|}$ (respectively) everywhere except at the origin. Here the singularities are not removable, and $\kappa \to \infty$ as we approach the origin. Moreover, $\kappa = 1$ everywhere on the one-dimensional interface in two spatial dimensions, and $\kappa = 2$ everywhere on the two-dimensional interface in three spatial dimensions. The difference occurs because a two-dimensional circle is a cylinder in three spatial dimensions (assuming that

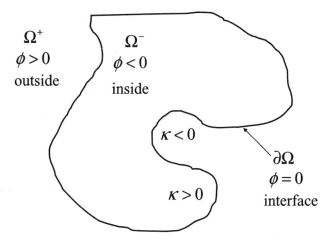

Figure 1.4. Convex regions have $\kappa > 0$, and concave regions have $\kappa < 0$.

the unmodeled direction has uniform data). It seems nonsensical to be troubled by $\kappa \to \infty$ as we approach the origin, since this is only a consequence of the embedding. In fact, since the smallest unit of measure on the Cartesian grid is the cell size $\triangle x$, it makes little sense to hope to resolve objects smaller than this. That is, it makes little sense to model circles (or spheres) with a radius smaller than $\triangle x$. Therefore, we limit the curvature so that $-\frac{1}{\triangle x} \leq \kappa \leq \frac{1}{\triangle x}$. If a value of κ is calculated outside this range, we merely replace that value with either $-\frac{1}{\triangle x}$ or $\frac{1}{\triangle x}$ depending on which is closer.

As a final note on curvature, one has to use caution when ϕ is noisy. The normal $\vec{N}$ will generally have even more noise, since it is based on the derivatives of ϕ. Similarly, the curvature κ will be even noisier than the normal, since it is computed with the second derivatives of ϕ.

1.5   Calculus Toolbox

The *characteristic function* χ^- of the interior region Ω^- is defined as

$$\chi^-(\vec{x}) = \begin{cases} 1 & \text{if } \phi(\vec{x}) \leq 0, \\ 0 & \text{if } \phi(\vec{x}) > 0 \end{cases} \tag{1.10}$$

where we arbitrarily include the boundary with the interior region. The characteristic function χ^+ of the exterior region Ω^+ is defined similarly as

$$\chi^+(\vec{x}) = \begin{cases} 0 & \text{if } \phi(\vec{x}) \leq 0, \\ 1 & \text{if } \phi(\vec{x}) > 0, \end{cases} \tag{1.11}$$

again including the boundary with the interior region. It is often useful to have only interior and exterior regions so that special treatment is not needed for the boundary. This is easily accomplished by including the measure-zero boundary set with either the interior or exterior region (as above). Throughout the text we usually include the boundary with the interior region Ω^- where $\phi(\vec{x}) < 0$ (unless otherwise specified).

The functions $\chi^\pm$ are functions of a multidimensional variable $\vec{x}$. It is often more convenient to work with functions of the one-dimensional variable ϕ. Thus we define the one-dimensional *Heaviside function*

$$H(\phi) = \begin{cases} 0 & \text{if } \phi \le 0, \\ 1 & \text{if } \phi > 0, \end{cases} \tag{1.12}$$

where ϕ depends on $\vec{x}$, although it is not important to specify this dependence when working with H. This allows us to work with H in one spatial dimension. Note that $\chi^+(\vec{x}) = H(\phi(\vec{x}))$ and $\chi^-(\vec{x}) = 1 - H(\phi(\vec{x}))$ for all $\vec{x}$, so all we have done is to introduce an extra function of one variable H to be used as a tool when dealing with characteristic functions.

The *volume integral* (area or length integral in $\Re^2$ or $\Re^1$, respectively) of a function f over the interior region Ω^- is defined as

$$\int_\Omega f(\vec{x})\chi^-(\vec{x}) \, d\vec{x}, \tag{1.13}$$

where the region of integration is all of Ω, since χ^- prunes out the exterior region Ω^+ automatically. The one-dimensional Heaviside function can be used to rewrite this volume integral as

$$\int_\Omega f(\vec{x}) \left(1 - H(\phi(\vec{x}))\right) \, d\vec{x} \tag{1.14}$$

representing the integral of f over the interior region Ω^-. Similarly,

$$\int_\Omega f(\vec{x}) H(\phi(\vec{x})) \, d\vec{x} \tag{1.15}$$

is the integral of f over the exterior region Ω^+.

By definition, the directional derivative of the Heaviside function H in the normal direction $\vec{N}$ is the *Dirac delta function*

$$\hat{\delta}(\vec{x}) = \nabla H(\phi(\vec{x})) \cdot \vec{N}, \tag{1.16}$$

which is a function of the multidimensional variable $\vec{x}$. Note that this distribution is nonzero only on the interface $\partial\Omega$ where $\phi = 0$. We can rewrite equation (1.16) as

$$\hat{\delta}(\vec{x}) = H'(\phi(\vec{x}))\nabla\phi(\vec{x}) \cdot \frac{\nabla\phi(\vec{x})}{|\nabla\phi(\vec{x})|} = H'(\phi(\vec{x}))|\nabla\phi(\vec{x})| \tag{1.17}$$

using the chain rule to take the gradient of H, the definition of the normal from equation (1.2), and the fact that $\nabla\phi(\vec{x}) \cdot \nabla\phi(\vec{x}) = |\nabla\phi(\vec{x})|^2$. In one spatial dimension, the *delta function* is defined as the derivative of the one-dimensional Heaviside function:

$$\delta(\phi) = H'(\phi), \tag{1.18}$$

where $H(\phi)$ is defined in equation (1.12) above. The delta function $\delta(\phi)$ is identically zero everywhere except at $\phi = 0$. This allows us to rewrite equations (1.16) and (1.17) as

$$\hat{\delta}(\vec{x}) = \delta(\phi(\vec{x}))|\nabla\phi(\vec{x})| \tag{1.19}$$

using the one-dimensional delta function $\delta(\phi)$.

The *surface integral* (line or point integral in $\Re^2$ or $\Re^1$, respectively) of a function f over the boundary $\partial\Omega$ is defined as

$$\int_\Omega f(\vec{x})\hat{\delta}(\vec{x})\,d\vec{x}, \tag{1.20}$$

where the region of integration is all of Ω, since $\hat{\delta}$ prunes out everything except $\partial\Omega$ automatically. The one-dimensional delta function can be used to rewrite this surface integral as

$$\int_\Omega f(\vec{x})\delta(\phi(\vec{x}))|\nabla\phi(\vec{x})|\,d\vec{x}. \tag{1.21}$$

Typically, volume integrals are computed by dividing up the interior region Ω^-, and surface integrals are computed by dividing up the boundary $\partial\Omega$. This requires treating a complex two-dimensional surface in three spatial dimensions. By embedding the volume and surface integrals in higher dimensions, equations (1.14), (1.15) and (1.21) avoid the need for identifying inside, outside, or boundary regions. Instead, the integrals are taken over the entire region Ω. Note that $d\vec{x}$ is a volume element in three spatial dimensions, an area element in two spatial dimensions, and a length element in one spatial dimension. On our Cartesian grid, the volume of a three-dimensional cell is $\triangle x\triangle y\triangle z$, the area of a two-dimensional cell is $\triangle x\triangle y$, and the length of a one-dimensional cell is $\triangle x$.

Consider the surface integral in equation (1.21), where the one-dimensional delta function $\delta(\phi)$ needs to be evaluated. Since $\delta(\phi) = 0$ almost everywhere, i.e., except on the lower-dimensional interface, which has measure zero, it seems unlikely that any standard numerical approximation based on sampling will give a good approximation to this integral. Thus, we use a first-order accurate smeared-out approximation of $\delta(\phi)$. First, we define the smeared-out Heaviside function

$$H(\phi) = \begin{cases} 0 & \phi < -\epsilon, \\ \frac{1}{2} + \frac{\phi}{2\epsilon} + \frac{1}{2\pi}\sin\left(\frac{\pi\phi}{\epsilon}\right) & -\epsilon \le \phi \le \epsilon, \\ 1 & \epsilon < \phi, \end{cases} \tag{1.22}$$

where ϵ is a tunable parameter that determines the size of the bandwidth of numerical smearing. A typically good value is $\epsilon = 1.5\triangle x$ (making the interface width equal to three grid cells when ϕ is normalized to a signed distance function with $|\nabla\phi| = 1$; see Chapter 2). Then the delta function is defined according to equation (1.18) as the derivative of the Heaviside function

$$\delta(\phi) = \begin{cases} 0 & \phi < -\epsilon, \\ \frac{1}{2\epsilon} + \frac{1}{2\epsilon}\cos\left(\frac{\pi\phi}{\epsilon}\right) & -\epsilon \le \phi \le \epsilon, \\ 0 & \epsilon < \phi, \end{cases} \tag{1.23}$$

where ϵ is determined as above. This delta function allows us to evaluate the surface integral in equation (1.21) using a standard sampling technique such as the midpoint rule. Similarly, the smeared-out Heaviside function in equation (1.22) allows us to evaluate the integrals in equations (1.14) and (1.15).

The reader is cautioned that the smeared-out Heaviside and delta functions approach to the calculus of implicit functions leads to first-order accurate methods. For example, when calculating the volume of the region Ω^- using

$$\int_{\Omega} (1 - H(\phi(\vec{x}))) \, dV \tag{1.24}$$

with the smeared-out Heaviside function in equation (1.22) (and $f(\vec{x}) = 1$), the errors in the calculation are $O(\triangle x)$ regardless of the accuracy of the integration method used. If one needs more accurate results, a three-dimensional contouring algorithm such as the *marching cubes* algorithm can be used to identify the region Ω^- more accurately, see Lorenson and Cline [108] or the more recent Kobbelt et al. [98]. Since higher-order accurate methods can be complex, we prefer the smeared-out Heaviside and delta function methods whenever appropriate.

2
Signed Distance Functions

2.1 Introduction

In the last chapter we defined implicit functions with $\phi(\vec{x}) \leq 0$ in the interior region Ω^-, $\phi(\vec{x}) > 0$ in the exterior region Ω^+, and $\phi(\vec{x}) = 0$ on the boundary $\partial\Omega$. Little was said about ϕ otherwise, except that smoothness is a desirable property especially in sampling the function or using numerical approximations. In this chapter we discuss signed distance functions, which are a subset of the implicit functions defined in the last chapter. We define signed distance functions to be positive on the exterior, negative on the interior, and zero on the boundary. An extra condition of $|\nabla\phi(\vec{x})| = 1$ is imposed on a signed distance function.

2.2 Distance Functions

A *distance function* $d(\vec{x})$ is defined as

$$d(\vec{x}) = \min(|\vec{x} - \vec{x}_I|) \quad \text{for all} \quad \vec{x}_I \in \partial\Omega, \tag{2.1}$$

implying that $d(\vec{x}) = 0$ on the boundary where $\vec{x} \in \partial\Omega$. Geometrically, d may be constructed as follows. If $\vec{x} \in \partial\Omega$, then $d(\vec{x}) = 0$. Otherwise, for a given point $\vec{x}$, find the point on the boundary set $\partial\Omega$ closest to $\vec{x}$, and label this point $\vec{x}_C$. Then $d(\vec{x}) = |\vec{x} - \vec{x}_C|$.

For a given point $\vec{x}$, suppose that $\vec{x}_C$ is the point on the interface closest to $\vec{x}$. Then for every point $\vec{y}$ on the line segment connecting $\vec{x}$ and $\vec{x}_C$,

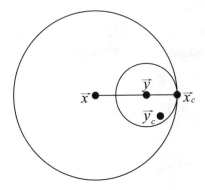

Figure 2.1. $\vec{x}_C$ is the closest interface point to $\vec{x}$ and $\vec{y}$.

$\vec{x}_C$ is the point on the interface closest to $\vec{y}$ as well. To see this, consider Figure 2.1, where $\vec{x}$, $\vec{x}_C$, and an example of a $\vec{y}$ are shown. Since $\vec{x}_C$ is the closest interface point to $\vec{x}$, no other interface points can be inside the large circle drawn about $\vec{x}$ passing through $\vec{x}_C$. Points closer to $\vec{y}$ than $\vec{x}_C$ must reside inside the small circle drawn about $\vec{y}$ passing through $\vec{x}_C$. Since the small circle lies inside the larger circle, no interface points can be inside the smaller circle, and thus $\vec{x}_C$ is the interface point closest to $\vec{y}$. The line segment from $\vec{x}$ to $\vec{x}_C$ is the shortest path from $\vec{x}$ to the interface. Any local deviation from this line segment increases the distance from the interface. In other words, the path from $\vec{x}$ to $\vec{x}_C$ is the path of steepest descent for the function d. Evaluating $-\nabla d$ at any point on the line segment from $\vec{x}$ to $\vec{x}_C$ gives a vector that points from $\vec{x}$ to $\vec{x}_C$. Furthermore, since d is Euclidean distance,

$$|\nabla d| = 1, \tag{2.2}$$

which is intuitive in the sense that moving twice as close to the interface gives a value of d that is half as big.

The above argument leading to equation (2.2) is true for any $\vec{x}$ as long as there is a unique closest point $\vec{x}_C$. That is, equation (2.2) is true except at points that are equidistant from (at least) two distinct points on the interface. Unfortunately, these equidistant points can exist, making equation (2.2) only generally true. It is also important to point out that equation (2.2) is generally only approximately satisfied in estimating the gradient numerically. One of the triumphs of the level set method involves the ease with which these degenerate points are treated numerically.

2.3   Signed Distance Functions

A *signed distance function* is an implicit function ϕ with $|\phi(\vec{x})| = d(\vec{x})$ for all $\vec{x}$. Thus, $\phi(\vec{x}) = d(\vec{x}) = 0$ for all $\vec{x} \in \partial\Omega$, $\phi(\vec{x}) = -d(\vec{x})$ for all $\vec{x} \in \Omega^-$, and $\phi(\vec{x}) = d(\vec{x})$ for all $\vec{x} \in \Omega^+$. Signed distance functions share all the properties of implicit functions discussed in the last chapter. In addition, there are a number of new properties that only signed distance functions possess. For example,

$$|\nabla \phi| = 1 \qquad (2.3)$$

as in equation (2.2). Once again, equation (2.3) is true only in a general sense. It is not true for points that are equidistant from at least two points on the interface. Distance functions have a kink at the interface where $d = 0$ is a local minimum, causing problems in approximating derivatives on or near the interface. On the other hand, signed distance functions are monotonic across the interface and can be differentiated there with significantly higher confidence.

Given a point $\vec{x}$, and using the fact that $\phi(\vec{x})$ is the signed distance to the closest point on the interface, we can write

$$\vec{x}_C = \vec{x} - \phi(\vec{x})\vec{N} \qquad (2.4)$$

to calculate the closet point on the interface, where $\vec{N}$ is the local unit normal at $\vec{x}$. Again, this is true only in a general sense, since equidistant points $\vec{x}$ have more than one closest point $\vec{x}_C$. Also, on our Cartesian grid, equation (2.4) will be only an approximation of the closest point on the interface $\vec{x}_C$. Nevertheless, we will find formulas of this sort very useful.

Equations that are true in a general sense can be used in numerical approximations as long as they fail in a graceful way that does not cause an overall deterioration of the numerical method. This is a general and powerful guideline for any numerical approach. So while the user should be cautiously knowledgeable of the possible failure of equations that are only generally true, one need not worry too much if the equation fails in a graceful (harmless) manner. More important, if the failure of an equation that is true in a general sense causes overall degradation of the numerical method, then many times a special-case approach can fix the problem. For example, when calculating the normals using equation (1.2) in the last chapter, we treated the special case where the denominator $|\nabla \phi|$ was identically zero by randomly choosing the normal direction. The numerical methods outlined in Part II of this book are based on vanishing viscosity solutions that guarantee reasonable behavior even at the occasional kink where a derivative fails to exist.

2.4 Examples

In the last chapter we used $\phi(x) = x^2 - 1$ as an implicit representation of $\partial \Omega = \{-1, 1\}$. A signed distance function representation of these same points is $\phi(x) = |x| - 1$, as shown in Figure 2.2. The signed distance function

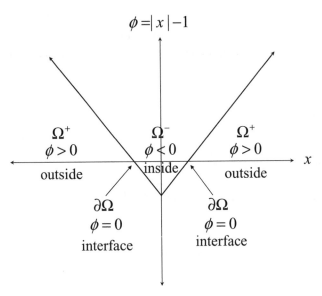

Figure 2.2. Signed distance function $\phi(x) = |x| - 1$ defining the regions Ω^- and Ω^+ as well as the boundary $\partial\Omega$.

$\phi(x) = |x| - 1$, gives the same boundary $\partial\Omega$, interior region Ω^-, and exterior region Ω^+, that the implicit function $\phi(x) = x^2 - 1$ did. However, the signed distance function $\phi(x) = |x| - 1$ has $|\nabla\phi| = 1$ for all $x \neq 0$. At $x = 0$ there is a kink in our function, and the derivative is not defined. While this may seem problematic, for example for determining the normal, our Cartesian grid contains only sample points and therefore cannot resolve this kink. On the Cartesian grid this kink is slightly smeared out, and the derivative will have a finite value. In fact, consideration of the possible placement of sample points shows that the value of the derivative lies in the interval $[-1, 1]$. Thus, nothing special needs to be done for kinks. In the worst-case scenario, the gradient vanishes at a kink, and remedies for this were already addressed in the last chapter.

In two spatial dimension we replace the implicit function $\phi(\vec{x}) = x^2 + y^2 - 1$ with the signed distance function $\phi(\vec{x}) = \sqrt{x^2 + y^2} - 1$ in order to implicitly represent the unit circle $\partial\Omega = \{\vec{x} \mid |\vec{x}| = 1\}$. Here $|\nabla\phi| = 1$ for all $x \neq 0$, and a multidimensional kink exists at the single point $x = 0$. Again, on our Cartesian grid the kink will be rounded out slightly and will not pose a problem. However, this numerical smearing of the kink makes $|\nabla\phi| \neq 1$ locally. That is, locally ϕ is no longer a signed distance function, and one has to take care when applying formulas that assume $|\nabla\phi| = 1$. Luckily, this does not generally lead to catastrophic difficulties. In fact, these kinks mostly exist away from the zero isocontour, which is the region of real interest in interface calculations.

In three spatial dimensions we replace the implicit function $\phi(\vec{x}) = x^2 + y^2 + z^2 - 1$ with the signed distance function $\phi(\vec{x}) = \sqrt{x^2 + y^2 + z^2} - 1$ in order to represent the surface of the unit sphere $\partial\Omega = \{\vec{x} \mid |\vec{x}| = 1\}$ implicitly. Again, the multidimensional kink at $x = 0$ will be smeared out on our Cartesian grid.

In all three examples there was a kink at a single point. This is somewhat misleading in general. For example, consider the one-dimensional example $\phi(x) = |x| - 1$ again, but in two spatial dimensions, where we write $\phi(\vec{x}) = |x| - 1$. Here, the interface consists of the two lines $x = -1$ and $x = 1$, and the interior region is $\Omega^- = \{\vec{x} \mid |x| < 1\}$. In this example every point along the line $x = 0$ has a kink in the x direction; i.e., there is an entire line of kinks. Similarly, in three spatial dimensions $\phi(\vec{x}) = |x| - 1$ implicitly represents the two planes $x = -1$ and $x = 1$. In this case every point on the two-dimensional plane $x = 0$ has a kink in the x direction; i.e., there is an entire plane of kinks. All of these kinks will be numerically smeared out on our Cartesian grid, and we need not worry about the derivative being undefined. However, locally $|\nabla\phi| \neq 1$ numerically.

2.5   Geometry and Calculus Toolboxes

Boolean operations for signed distance functions are similar to those for general implicit functions. If ϕ_1 and ϕ_2 are two different signed distance functions, then $\phi(\vec{x}) = \min(\phi_1(\vec{x}), \phi_2(\vec{x}))$ is the signed distance function representing the union of the interior regions. The function $\phi(\vec{x}) = \max(\phi_1(\vec{x}), \phi_2(\vec{x}))$ is the signed distance function, representing the intersection of the interior regions. The complement of the set defined by $\phi_1(\vec{x})$ has signed distance function $\phi(\vec{x}) = -\phi_1(\vec{x})$. Also, $\phi(\vec{x}) = \max(\phi_1(\vec{x}), -\phi_2(\vec{x}))$ is the signed distance function for the region defined by subtracting the interior of ϕ_2 from the interior of ϕ_1.

As mentioned in the last chapter, we would like our implicit function to be as smooth as possible. It turns out that signed distance functions, especially those where the kinks have been numerically smeared, are probably the best candidates for implicit representation of interfaces. This is because $|\nabla\phi| = 1$ everywhere except near the smoothed-out kinks. This simplifies many of the formulas from the last chapter by removing the normalization constants. Equation (1.2) simplifies to

$$\vec{N} = \nabla\phi \tag{2.5}$$

for the local unit normal. Equation (1.7) simplifies to

$$\kappa = \triangle\phi \tag{2.6}$$

for the curvature, where $\triangle\phi$ is the Laplacian of ϕ defined as

$$\triangle\phi = \phi_{xx} + \phi_{yy} + \phi_{zz}, \tag{2.7}$$

which should not be confused with $\triangle x$, which is the size of a Cartesian grid cell. While this overuse of notation may seem confusing at first, it is very common and usually clarified from the context in which it is used.

Note the simplicity of equation (2.7) as compared to equation (1.8). Obviously, there is much to be gained in simplicity and efficiency in using signed distance functions. However, one should be relatively cautious, since smeared-out kinks will generally have $|\nabla\phi| \neq 1$, so that equations (2.5) and (2.6) do not accurately define the normal and the curvature. In fact, when using numerical approximations, one will not generally obtain $|\nabla\phi| = 1$, so equations (2.5) and (2.6) will not generally be accurate. There are many instances of the normal or the curvature appearing in a set of equations when these quantities may not actually be needed or desired. In fact, one may actually prefer the gradient of ϕ (i.e., $\nabla\phi$) instead of the normal. Similarly, one may prefer the Laplacian of ϕ (i.e., $\triangle\phi$) instead of the curvature. In this sense one should always keep equations (2.5) and (2.6) in mind when performing numerical calculations. Even if they are not generally true, they have the potential to make the calculations more efficient and even better behaved in some situations.

The multidimensional delta function in equation (1.19) can be rewritten as

$$\hat{\delta}(\vec{x}) = \delta(\phi(\vec{x})) \tag{2.8}$$

using the one-dimensional delta function $\delta(\phi)$. The surface integral in equation (1.21) then becomes

$$\int_{\Omega} f(\vec{x})\delta(\phi(\vec{x}))d\vec{x}, \tag{2.9}$$

where the $|\nabla\phi|$ term has been omitted.

Part II
Level Set Methods

Level set methods add *dynamics* to implicit surfaces. The key idea that started the level set fanfare was the Hamilton-Jacobi approach to numerical solutions of a time-dependent equation for a moving implicit surface. This was first done in the seminal work of Osher and Sethian [126]. In the following chapters we will discuss this seminal work along with many of the auxiliary equations that were developed along the way, including a general numerical approach for Hamilton-Jacobi equations.

In the first chapter we discuss the basic convection equation, otherwise known as the "level set equation." This moves an implicit surface in an externally generated velocity field. In the following chapter we discuss motion by mean curvature, emphasizing the parabolic nature of this equation as opposed to the underlying hyperbolic nature of the level set equation. Then, in the following chapter we introduce the general concept of Hamilton-Jacobi equations, noting that basic convection is a simple instance of this general framework. In the next chapter we discuss the concept of a surface moving normal to itself. The next two chapters address two of the core level set equations and give details for obtaining numerical solutions in the Hamilton-Jacobi framework. Specifically, we discuss reinitialization to a signed distance function and extrapolation of a quantity away from or across an interface. After this, we discuss a recently developed particle level set method that hybridizes the Eulerian level set approach with Lagrangian particle-tracking technology. Finally, we wrap up this part of the book with a brief discussion of codimension-two (and higher) objects.

3

Motion in an Externally Generated Velocity Field

3.1 Convection

Suppose that the velocity of each point on the implicit surface is given as $\vec{V}(\vec{x})$; i.e., assume that $\vec{V}(\vec{x})$ is known for every point $\vec{x}$ with $\phi(\vec{x}) = 0$. Given this velocity field $\vec{V} = \langle u, v, w \rangle$, we wish to move all the points on the surface with this velocity. The simplest way to do this is to solve the ordinary differential equation (ODE)

$$\frac{d\vec{x}}{dt} = \vec{V}(\vec{x}) \tag{3.1}$$

for every point $\vec{x}$ on the front, i.e., for all $\vec{x}$ with $\phi(\vec{x}) = 0$. This is the *Lagrangian* formulation of the interface evolution equation. Since there are generally an infinite number of points on the front (except, of course, in one spatial dimension), this means discretizing the front into a finite number of pieces. For example, one could use segments in two spatial dimensions or triangles in three spatial dimensions and move the endpoints of these segments or triangles. This is not so hard to accomplish if the connectivity does not change and the surface elements are not distorted too much. Unfortunately, even the most trivial velocity fields can cause large distortion of boundary elements (segments or triangles), and the accuracy of the method can deteriorate quickly if one does not periodically modify the discretization in order to account for these deformations by smoothing and regularizing inaccurate surface elements. The interested reader is referred to [174] for a rather recent least-squares-based smoothing scheme for damping mesh-instabilities due to deforming elements. Examples are given in both two and three spatial dimensions. Reference [174] also discusses the use of a mesh-refinement procedure to maintain some degree of regularity as the interface deforms. Again, without these special procedures for maintaining both smoothness and regularity, the interface can deteriorate to the point where numerical results are so inaccurate as to be unusable. In addition to dealing with element deformations, one must decide how to modify the interface discretization when the topology changes. These surgical methods of detaching and reattaching boundary elements can quickly become rather complicated. Reference [174] outlines some of the details involved in a single "surgical cut" of a three-dimensional surface. The use of the Lagrangian formulation of the interface motion given in equation (3.1) along with numerical techniques for smoothing, regularization, and surgery are collectively referred to as *front tracking* methods. A seminal work in the field of three-dimensional front tracking is [168], and the interested reader is referred to [165] for a current state-of-the-art review.

In order to avoid problems with instabilities, deformation of surface elements, and complicated surgical procedures for topological repair of interfaces, we use our implicit function ϕ both to represent the interface and to *evolve* the interface. In order to define the evolution of our implicit function ϕ we use the simple convection (or advection) equation

$$\phi_t + \vec{V} \cdot \nabla \phi = 0, \tag{3.2}$$

where the t subscript denotes a temporal partial derivative in the time variable t. Recall that ∇ is the gradient operator, so that

$$\vec{V} \cdot \nabla \phi = u\phi_x + v\phi_y + w\phi_z.$$

This partial differential equation (PDE) defines the motion of the interface where $\phi(\vec{x}) = 0$. It is an *Eulerian* formulation of the interface evolution, since the interface is *captured* by the implicit function ϕ as opposed to being *tracked* by interface elements as was done in the Lagrangian formulation. Equation (3.2) is sometimes referred to as the *level set equation*; it was introduced for numerical interface evolution by Osher and Sethian [126]. It is also a quite popular equation in the combustion community, where it is known as the *G-equation* given by

$$G_t + \vec{V} \cdot \nabla G = 0, \tag{3.3}$$

where the $G(\vec{x}) = 0$ isocontour is used to represent implicitly the reaction surface of an evolving flame front. The G-equation was introduced by Markstein [110], and it is used in the asymptotic analysis of flame fronts in instances where the front is thin enough to be considered a discontinuity. The interested reader is referred to Williams [173] as well. Lately, numerical practitioners in the combustion community have started using level set methods to find numerical solutions of equation (3.3) in (obviously) the same manner as equation (3.2).

On a Cartesian grid it can be slightly complicated to implement equation (3.2) if the velocity field is defined only on the interface itself. So, as with the embedding of ϕ, we usually write equation (3.2) using the assumption that the velocity field $\vec{V}$ is not only defined on the interface where $\phi(\vec{x}) = 0$, but is defined off the interface as well. Often $\vec{V}$ will be naturally defined on the entire computational domain Ω, but for numerical purposes it is usually sufficient to have $\vec{V}$ defined on a band containing the interface. The bandwidth varies based on the numerical method used to obtain approximate solutions to equation (3.2). When $\vec{V}$ is already defined throughout all of Ω nothing special need be done. However, there are interesting examples where $\vec{V}$ is known only on the interface, and one must extend its definition to (at least) a band about the interface in order to solve equation (3.2). We will discuss the extension of a velocity off the interface in Chapter 8.

Embedding $\vec{V}$ on our Cartesian grid introduces the same sampling issues that we faced in Chapter 1 when we embedded the interface Γ as the zero level set of the function ϕ. For example, suppose we were given a velocity field $\vec{V}$ that is identically zero in all of Ω except on the interface, where $\vec{V} = \langle 1, 0, 0 \rangle$. Then the exact solution is an interface moving to the right with speed 1. However, since most (if not all) of the Cartesian grid points will not lie on the interface, most of the points on our Cartesian mesh have $\vec{V}$ identically equal to zero, causing the $\vec{V} \cdot \nabla\phi$ term in equation (3.2) to vanish. This in turn implies that $\phi_t = 0$ almost everywhere, so that the interface mostly (or completely if no points happen to fall on the interface) incorrectly sits still. This difficult issue can be rectified in part by placing some conditions on the velocity field $\vec{V}$. For example, if we require that $\vec{V}$ be continuous near the interface, then this rules out our degenerate example.

Restricting $\vec{V}$ to the set of continuous functions generally does not alleviate our sampling problems. Suppose, for example, that the above velocity field was equal to $\langle 1, 0, 0 \rangle$ on the interface, zero outside a band of thickness $\epsilon > 0$ surrounding the interface, and smooth in between. We can choose $\vec{V}$ as smooth as we like by defining it appropriately in the band of thickness ϵ surrounding the interface. The difficulty arises when ϵ is small compared to $\triangle x$. If ϵ is small enough, then almost every grid point will lie outside the band where $\vec{V} = 0$. Once again, we will (mostly) compute an interface that incorrectly sits still. In fact, even if ϵ is comparable to $\triangle x$, the numerical solution will have significant errors. In order to *resolve* the velocity field, it is necessary to have a number of grid points within the ϵ thickness band surrounding the interface. That is, we need $\triangle x$ to be significantly smaller than the velocity variation (which scales like ϵ) in order get a good approximation of the velocity near the interface. Since $\triangle x$ needs to be much smaller than ϵ, we desire a relatively large ϵ to minimize the variation in the velocity field.

Given a velocity field $\vec{V}$ and the notion (discussed above) that minimizing its variation is good for treating the sampling problem, there is an obvious choice of $\vec{V}$ that gives both the correct interface motion and the least variation. First, since the values of $\vec{V}$ given on the interface dictate the correct interface motion, these cannot be changed, regardless of the variation. In some sense, the spatial variation of the velocity on the interface dictates how many Cartesian grid points will be needed to accurately predict the interface motion. If we cannot resolve the tangential variation of the interface velocity with our Cartesian grid, then it is unlikely that we can calculate a good approximation to the interface motion. Second, the velocity off the interface has nothing to do with the correct interface motion. This is true even if the velocity off the interface is inherited from some underlying physical calculation. Only the velocity of the interface itself contains any real information about the interface propagation. Otherwise, one would have no hope of using the Lagrangian formulation, equation (3.1), to calculate the interface motion. In summary, the velocity variation tangential to the interface dictates the interface motion, while the velocity variation normal to the interface is meaningless. Therefore, the minimum variation in the velocity field can be obtained by restricting the interface velocity $\vec{V}$ to be constant in the direction normal to the interface. This generally makes the velocity multivalued, since lines normal to the interface will eventually intersect somewhere away from the interface (if the interface has a nonzero curvature). Alternatively, the velocity $\vec{V}(\vec{x})$ at a point $\vec{x}$ can be set equal to the interface velocity $\vec{V}(\vec{x}_C)$ at the interface point $\vec{x}_C$ closest to the point $\vec{x}$. While this doesn't change the value of the velocity on the interface, it makes the velocity off the interface approximately constant in the normal direction local to the interface. In Chapter 8 we will discuss numerical techniques for constructing a velocity field defined in this manner.

Defining the velocity $\vec{V}$ equal to the interface velocity at the closest interface point $\vec{x}_C$ is a rather ingenious idea. In the appendix of [175], Zhao et al. showed that a signed distance function tends to stay a signed distance function if this closest interface point velocity is used to advect the interface. A number of researchers have been using this specially defined velocity field because it usually gives superior results over velocity fields with needlessly more spatial variation. Chen, Merriman, Osher, and Smereka [43] published the first numerical results based on this specially designed velocity field. The interested reader is referred to the rather interesting work of Adalsteinsson and Sethian [1] as well.

The velocity field given in equation (3.2) can come from a number of external sources. For example, when the $\phi(\vec{x}) = 0$ isocontour represents the interface between two different fluids, the interface velocity is calculated using the two-phase Navier-Stokes equations. This illustrates that the interface velocity more generally depends on both space and time and should be written as $\vec{V}(\vec{x}, t)$, but we occasionally omit the $\vec{x}$ dependence and more often the t dependence for brevity.

3.2   Upwind Differencing

Once ϕ and $\vec{V}$ are defined at every grid point (or at least sufficiently close to the interface) on our Cartesian grid, we can apply numerical methods to evolve ϕ forward in time moving the interface across the grid. At some point in time, say time t^n, let $\phi^n = \phi(t^n)$ represent the current values of ϕ. Updating ϕ in time consists of finding new values of ϕ at every grid point after some time increment $\triangle t$. We denote these new values of ϕ by $\phi^{n+1} = \phi(t^{n+1})$, where $t^{n+1} = t^n + \triangle t$.

A rather simple first-order accurate method for the time discretization of equation (3.2) is the *forward Euler* method given by

$$\frac{\phi^{n+1} - \phi^n}{\triangle t} + \vec{V}^n \cdot \nabla \phi^n = 0, \tag{3.4}$$

where $\vec{V}^n$ is the given external velocity field at time t^n, and $\nabla \phi^n$ evaluates the gradient operator using the values of ϕ at time t^n. Naively, one might evaluate the spatial derivatives of ϕ in a straightforward manner using equation (1.3), (1.4), or (1.5). Unfortunately, this straightforward approach will fail. One generally needs to exercise great care when numerically discretizing partial differential equations. We begin by writing equation (3.4) in expanded form as

$$\frac{\phi^{n+1} - \phi^n}{\triangle t} + u^n \phi_x^n + v^n \phi_y^n + w^n \phi_z^n = 0 \tag{3.5}$$

and address the evaluation of the $u^n \phi_x^n$ term first. The techniques used to approximate this term can then be applied independently to the $v^n \phi_y^n$ and $w^n \phi_z^n$ terms in a *dimension-by-dimension* manner.

For simplicity, consider the one-dimensional version of equation (3.5),

$$\frac{\phi^{n+1} - \phi^n}{\triangle t} + u^n \phi_x^n = 0, \tag{3.6}$$

where the sign of u^n indicates whether the values of ϕ are moving to the right or to the left. Since u^n can be spatially varying, we focus on a specific grid point x_i, where we write

$$\frac{\phi_i^{n+1} - \phi_i^n}{\triangle t} + u_i^n (\phi_x)_i^n = 0, \tag{3.7}$$

where $(\phi_x)_i$ denotes the spatial derivative of ϕ at the point x_i. If $u_i > 0$, the values of ϕ are moving from left to right, and the *method of characteristics* tells us to look to the left of x_i to determine what value of ϕ will land on the point x_i at the end of a time step. Similarly, if $u_i < 0$, the values of ϕ are moving from right to left, and the method of characteristics implies that we should look to the right to determine an appropriate value of ϕ_i at time t^{n+1}. Clearly, $D^- \phi$ (from equation (1.4)) should be used to approximate ϕ_x when $u_i > 0$. In contrast, $D^+ \phi$ cannot possibly give a good approximation, since it fails to contain the information to the left of x_i that dictates the new value of ϕ_i. Similar reasoning indicates that $D^+\phi$ should be used to approximate ϕ_x when $u_i < 0$. This method of choosing an approximation to the spatial derivatives based on the sign of u is known as upwind differencing or *upwinding*. Generally, upwind methods approximate derivatives by biasing the finite difference stencil in the direction where the characteristic information is coming from.

We summarize the upwind discretization as follows. At each grid point, define ϕ_x^- as $D^-\phi$ and ϕ_x^+ as $D^+\phi$. If $u_i > 0$, approximate ϕ_x with ϕ_x^-. If $u_i < 0$, approximate ϕ_x with ϕ_x^+. When $u_i = 0$, the $u_i(\phi_x)_i$ term vanishes, and ϕ_x does not need to be approximated. This is a first-order accurate discretization of the spatial operator, since $D^-\phi$ and $D^+\phi$ are first-order accurate approximations of the derivative; i.e., the errors are $O(\triangle x)$.

The combination of the forward Euler time discretization with the upwind difference scheme is a *consistent* finite difference approximation to the partial differential equation (3.2), since the approximation error converges to zero as $\triangle t \to 0$ and $\triangle x \to 0$. According to the Lax-Richtmyer equivalence theorem a finite difference approximation to a linear partial differential equation is *convergent*, i.e., the correct solution is obtained as $\triangle t \to 0$ and $\triangle x \to 0$, if and only if it is both consistent and *stable*. Stability guarantees that small errors in the approximation are not amplified as the solution is marched forward in time.

Stability can be enforced using the Courant-Friedreichs-Lewy condition (CFL condition), which asserts that the numerical waves should propagate at least as fast as the physical waves. This means that the numerical wave speed of $\triangle x/\triangle t$ must be at least as fast as the physical wave speed $|u|$, i.e., $\triangle x/\triangle t > |u|$. This leads us to the CFL *time step restriction* of

$$\triangle t < \frac{\triangle x}{\max\{|u|\}}, \tag{3.8}$$

where $\max\{|u|\}$ is chosen to be the largest value of $|u|$ over the entire Cartesian grid. In reality, we only need to choose the largest value of $|u|$ on the interface. Of course, these two values are the same if the velocity field is defined as the velocity of the closest point on the interface. Equation (3.8) is usually enforced by choosing a *CFL number* α with

$$\triangle t \left(\frac{\max\{|u|\}}{\triangle x}\right) = \alpha \tag{3.9}$$

and $0 < \alpha < 1$. A common near-optimal choice is $\alpha = 0.9$, and a common conservative choice is $\alpha = 0.5$. A multidimensional CFL condition can be written as

$$\triangle t \max\left\{\frac{|u|}{\triangle x} + \frac{|v|}{\triangle y} + \frac{|w|}{\triangle z}\right\} = \alpha, \tag{3.10}$$

although

$$\Delta t \left(\frac{\max\{|\vec{V}|\}}{\min\{\Delta x, \Delta y, \Delta z\}} \right) = \alpha \tag{3.11}$$

is also quite popular. More details on consistency, stability, and convergence can be found in basic textbooks on the numerical solution of partial differential equations; see, for example, [157].

Instead of upwinding, the spatial derivatives in equation (3.2) could be approximated with the more accurate central differencing. Unfortunately, simple central differencing is unstable with forward Euler time discretization and the usual CFL conditions with $\Delta t \sim \Delta x$. Stability can be achieved by using a much more restrictive CFL condition with $\Delta t \sim (\Delta x)^2$, although this is too computationally costly. Stability can also be achieved by using a different temporal discretization, e.g., the third-order accurate Runge-Kutta method (discussed below). A third way of achieving stability consists in adding some artificial dissipation to the right-hand side of equation (3.2) to obtain

$$\phi_t + \vec{V} \cdot \nabla\phi = \mu\Delta\phi, \tag{3.12}$$

where the viscosity coefficient μ is chosen proportional to Δx, i.e., $\mu \sim \Delta x$, so that the *artificial viscosity* vanishes as $\Delta x \to 0$, enforcing consistency for this method. While all three of these approaches stabilize central differencing, we instead prefer to use upwind methods, which draw on the highly sucessful technology developed for the numerical solution of conservation laws.

3.3   Hamilton-Jacobi ENO

The first-order accurate upwind scheme described in the last section can be improved upon by using a more accurate approximation for ϕ_x^- and ϕ_x^+. The velocity u is still used to decide whether ϕ_x^- or ϕ_x^+ is used, but the approximations for ϕ_x^- or ϕ_x^+ can be improved significantly.

In [81], Harten et al. introduced the idea of essentially nonoscillatory (ENO) polynomial interpolation of data for the numerical solution of conservation laws. Their basic idea was to compute numerical flux functions using the smoothest possible polynomial interpolants. The actual numerical implementation of this idea was improved considerably by Shu and Osher in [150] and [151], where the numerical flux functions were constructed directly from a divided difference table of the pointwise data. In [126], Osher and Sethian realized that Hamilton-Jacobi equations in one spatial dimension are integrals of conservation laws. They used this fact to extend the ENO method for the numerical discretization of conservation laws to Hamilton-Jacobi equations such as equation (3.2). This Hamilton-Jacobi

ENO (HJ ENO) method allows one to extend first-order accurate upwind differencing to higher-order spatial accuracy by providing better numerical approximations to ϕ_x^- or ϕ_x^+.

Proceeding along the lines of [150] and [151], we use the smoothest possible polynomial interpolation to find ϕ and then differentiate to get ϕ_x. As is standard with Newton polynomial interpolation (see any undergraduate numerical analysis text, e.g., [82]), the zeroth divided differences of ϕ are defined at the grid nodes and defined by

$$D_i^0 \phi = \phi_i \tag{3.13}$$

at each grid node i (located at x_i). The first divided differences of ϕ are defined midway between grid nodes as

$$D_{i+1/2}^1 \phi = \frac{D_{i+1}^0 \phi - D_i^0 \phi}{\triangle x}, \tag{3.14}$$

where we are assuming that the mesh spacing is uniformly $\triangle x$. Note that $D_{i-1/2}^1 \phi = (D^- \phi)_i$ and $D_{i+1/2}^1 \phi = (D^+ \phi)_i$, i.e., the first divided differences, are the backward and forward difference approximations to the derivatives. The second divided differences are defined at the grid nodes as

$$D_i^2 \phi = \frac{D_{i+1/2}^1 \phi - D_{i-1/2}^1 \phi}{2\triangle x}, \tag{3.15}$$

while the third divided differences

$$D_{i+1/2}^3 \phi = \frac{D_{i+1}^2 \phi - D_i^2 \phi}{3\triangle x} \tag{3.16}$$

are defined midway between the grid nodes.

The divided differences are used to reconstruct a polynomial of the form

$$\phi(x) = Q_0(x) + Q_1(x) + Q_2(x) + Q_3(x) \tag{3.17}$$

that can be differentiated and evaluated at x_i to find $(\phi_x^-)_i$ and $(\phi_x^+)_i$. That is, we use

$$\phi_x(x_i) = Q_1'(x_i) + Q_2'(x_i) + Q_3'(x_i) \tag{3.18}$$

to define $(\phi_x^-)_i$ and $(\phi_x^+)_i$. Note that the constant $Q_0(x)$ term vanishes upon differentiation.

To find ϕ_x^- we start with $k = i - 1$, and to find ϕ_x^+ we start with $k = i$. Then we define

$$Q_1(x) = (D_{k+1/2}^1 \phi)(x - x_i), \tag{3.19}$$

so that

$$Q_1'(x_i) = D_{k+\frac{1}{2}}^1 \phi, \tag{3.20}$$

implying that the contribution from $Q_1'(x_i)$ in equation (3.18) is the backward difference in the case of ϕ_x^- and the forward difference in the case of ϕ_x^+. In other words, first-order accurate polynomial interpolation is exactly first-order upwinding. Improvements are obtained by including the $Q_2'(x_i)$ and $Q_3'(x_i)$ terms in equation (3.18), leading to second- and third-order accuracy, respectively.

Looking at the divided difference table and noting that $D_{k+1/2}^1\phi$ was chosen for first-order accuracy, we have two choices for the second-order accurate correction. We could include the next point to the left and use $D_k^2\phi$, or we could include the next point to the right and use $D_{k+1}^2\phi$. The key observation is that smooth slowly varying data tend to produce small numbers in divided difference tables, while discontinuous or quickly varying data tend to produce large numbers in divided difference tables. This is obvious in the sense that the differences measure variation in the data. Comparing $|D_k^2\phi|$ to $|D_{k+1}^2\phi|$ indicates which of the polynomial interpolants has more variation. We would like to avoid interpolating near large variations such as discontinuities or steep gradients, since they cause overshoots in the interpolating function, leading to numerical errors in the approximation of the derivative. Thus, if $|D_k^2\phi| \le |D_{k+1}^2\phi|$, we set $c = D_k^2\phi$ and $k^\star = k - 1$; otherwise, we set $c = D_{k+1}^2\phi$ and $k^\star = k$. Then we define

$$Q_2(x) = c(x - x_k)(x - x_{k+1}), \tag{3.21}$$

so that

$$Q_2'(x_i) = c\,(2(i - k) - 1)\,\triangle x \tag{3.22}$$

is the second-order accurate correction to the approximation of ϕ_x in equation (3.18). If we stop here, i.e., omitting the Q_3 term, we have a second-order accurate method for approximating ϕ_x^- and ϕ_x^+. Note that $k^\star$ has not yet been used. It is defined below for use in calculating the third-order accurate correction.

Similar to the second-order accurate correction, the third-order accurate correction is obtained by comparing $|D_{k^\star+1/2}^3\phi|$ and $|D_{k^\star+3/2}^3\phi|$. If $|D_{k^\star+1/2}^3\phi| \le |D_{k^\star+3/2}^3\phi|$, we set $c^\star = D_{k^\star+1/2}^3\phi$; otherwise, we set $c^\star = D_{k^\star+3/2}^3\phi$. Then we define

$$Q_3(x) = c^\star(x - x_{k^\star})(x - x_{k^\star+1})(x - x_{k^\star+2}), \tag{3.23}$$

so that

$$Q_3'(x_i) = c^\star\left(3(i - k^\star)^2 - 6(i - k^\star) + 2\right)(\triangle x)^2 \tag{3.24}$$

is the third-order accurate correction to the approximation of ϕ_x in equation (3.18).

3.4   Hamilton-Jacobi WENO

When calculating $(\phi_x^-)_i$, the third-order accurate HJ ENO scheme uses a subset of $\{\phi_{i-3}, \phi_{i-2}, \phi_{i-1}, \phi_i, \phi_{i+1}, \phi_{i+2}\}$ that depends on how the stencil is chosen. In fact, there are exactly three possible HJ ENO approximations to $(\phi_x^-)_i$. Defining $v_1 = D^-\phi_{i-2}$, $v_2 = D^-\phi_{i-1}$, $v_3 = D^-\phi_i$, $v_4 = D^-\phi_{i+1}$, and $v_5 = D^-\phi_{i+2}$ allows us to write

$$\phi_x^1 = \frac{v_1}{3} - \frac{7v_2}{6} + \frac{11v_3}{6}, \tag{3.25}$$

$$\phi_x^2 = -\frac{v_2}{6} + \frac{5v_3}{6} + \frac{v_4}{3}, \tag{3.26}$$

and

$$\phi_x^3 = \frac{v_3}{3} + \frac{5v_4}{6} - \frac{v_5}{6} \tag{3.27}$$

as the three potential HJ ENO approximations to ϕ_x^-. The goal of HJ ENO is to choose the single approximation with the least error by choosing the smoothest possible polynomial interpolation of ϕ.

In [107], Liu et al. pointed out that the ENO philosophy of picking exactly one of three candidate stencils is overkill in smooth regions where the data are well behaved. They proposed a weighted ENO (WENO) method that takes a convex combination of the three ENO approximations. Of course, if any of the three approximations interpolates across a discontinuity, it is given minimal weight in the convex combination in order to minimize its contribution and the resulting errors. Otherwise, in smooth regions of the flow, all three approximations are allowed to make a significant contribution in a way that improves the local accuracy from third order to fourth order. Later, Jiang and Shu [89] improved the WENO method by choosing the convex combination weights in order to obtain the optimal fifth-order accuracy in smooth regions of the flow. In [88], following the work on HJ ENO in [127], Jiang and Peng extended WENO to the Hamilton-Jacobi framework. This Hamilton-Jacobi WENO, or HJ WENO, scheme turns out to be very useful for solving equation (3.2), since it reduces the errors by more than an order of magnitude over the third-order accurate HJ ENO scheme for typical applications.

The HJ WENO approximation of $(\phi_x^-)_i$ is a convex combination of the approximations in equations (3.25), (3.26), and (3.27) given by

$$\phi_x = \omega_1\phi_x^1 + \omega_2\phi_x^2 + \omega_3\phi_x^3, \tag{3.28}$$

where the $0 \leq \omega_k \leq 1$ are the weights with $\omega_1 + \omega_2 + \omega_3 = 1$. The key observation for obtaining high-order accuracy in smooth regions is that weights of $\omega_1 = 0.1$, $\omega_2 = 0.6$ and $\omega_3 = 0.3$ give the optimal fifth-order accurate approximation to ϕ_x. While this is the optimal approximation, it is valid only in smooth regions. In nonsmooth regions this optimal weighting can be very inaccurate, and we are better off with digital ($\omega_k = 0$ or $\omega_k = 1$) weights that choose a single approximation to ϕ_x, i.e., the HJ ENO approximation.

Reference [89] pointed out that setting $\omega_1 = 0.1 + O((\triangle x)^2)$, $\omega_2 = 0.6 + O((\triangle x)^2)$, and $\omega_3 = 0.3 + O((\triangle x)^2)$ still gives the optimal fifth-order accuracy in smooth regions. In order to see this, we rewrite these as $\omega_1 = 0.1 + C_1(\triangle x)^2$, $\omega_2 = 0.6 + C_2(\triangle x)^2$ and $\omega_3 = 0.3 + C_3(\triangle x)^2$ and plug them into equation (3.28) to obtain

$$0.1\phi_x^1 + 0.6\phi_x^2 + 0.3\phi_x^3 \tag{3.29}$$

and

$$C_1(\triangle x)^2\phi_x^1 + C_2(\triangle x)^2\phi_x^2 + C_3(\triangle x)^2\phi_x^3 \tag{3.30}$$

as the two terms that are added to give the HJ WENO approximation to ϕ_x. The term given by equation (3.29) is the optimal approximation that gives the exact value of ϕ_x plus an $O((\triangle x)^5)$ error term. Thus, if the term given by equation (3.30) is $O((\triangle x)^5)$, then the entire HJ WENO approximation is $O((\triangle x)^5)$ in smooth regions. To see that this is the case, first note that each of the HJ ENO ϕ_x^k approximations gives the exact value of ϕ_x, denoted by ϕ_x^E, plus an $O((\triangle x)^3)$ error term (in smooth regions). Thus, the term in equation (3.30) is

$$C_1(\triangle x)^2\phi_x^E + C_2(\triangle x)^2\phi_x^E + C_3(\triangle x)^2\phi_x^E \tag{3.31}$$

plus an $O((\triangle x)^2)O((\triangle x)^3) = O((\triangle x)^5)$ term. Since, each of the C_k is $O(1)$, as is ϕ_x^E, this appears to be an $O((\triangle x)^2)$ term at first glance. However, since $\omega_1 + \omega_2 + \omega_3 = 1$, we have $C_1 + C_2 + C_3 = 0$, implying that the term in equation (3.31) is identically zero. Thus, the HJ WENO approximation is $O((\triangle x)^5)$ in smooth regions. Note that [107] obtained only fourth-order accuracy, since they chose $\omega_1 = 0.1 + O(\triangle x)$, $\omega_2 = 0.6 + O(\triangle x)$, and $\omega_3 = 0.3 + O(\triangle x)$.

In order to define the weights, ω_k, we follow [88] and estimate the smoothness of the stencils in equations (3.25), (3.26), and (3.27) as

$$S_1 = \frac{13}{12}(v_1 - 2v_2 + v_3)^2 + \frac{1}{4}(v_1 - 4v_2 + 3v_3)^2, \tag{3.32}$$

$$S_2 = \frac{13}{12}(v_2 - 2v_3 + v_4)^2 + \frac{1}{4}(v_2 - v_4)^2, \tag{3.33}$$

and

$$S_3 = \frac{13}{12}(v_3 - 2v_4 + v_5)^2 + \frac{1}{4}(3v_3 - 4v_4 + v_5)^2, \tag{3.34}$$

respectively. Using these smoothness estimates, we define

$$\alpha_1 = \frac{0.1}{(S_1 + \epsilon)^2}, \tag{3.35}$$

$$\alpha_2 = \frac{0.6}{(S_2 + \epsilon)^2}, \tag{3.36}$$

and

$$\alpha_3 = \frac{.3}{(S_3 + \epsilon)^2} \tag{3.37}$$

with

$$\epsilon = 10^{-6} \max \left\{ v_1^2, v_2^2, v_3^2, v_4^2, v_5^2 \right\} + 10^{-99}, \tag{3.38}$$

where the 10^{-99} term is set to avoid division by zero in the definition of the α_k. This value for epsilon was first proposed by Fedkiw et al. [69], where the first term is a scaling term that aids in the balance between the optimal fifth-order accurate stencil and the digital HJ ENO weights. In the case that ϕ is an approximate signed distance function, the v_k that approximate ϕ_x are approximately equal to one, so that the first term in equation (3.38) can be set to 10^{-6}. This first term can then absorb the second term, yielding $\epsilon = 10^{-6}$ in place of equation (3.38). Since the first term in equation (3.38) is only a scaling term, it is valid to make this $v_k \approx 1$ estimate in higher dimensions as well.

A smooth solution has small variation leading to small S_k. If the S_k are small enough compared to ϵ, then equations (3.35), (3.36), and (3.37) become $\alpha_1 \approx 0.1\epsilon^{-2}$, $\alpha_2 \approx 0.6\epsilon^{-2}$, and $\alpha_3 \approx 0.3\epsilon^{-2}$, exhibiting the proper ratios for the optimal fifth-order accuracy. That is, normalizing the α_k to obtain the weights

$$\omega_1 = \frac{\alpha_1}{\alpha_1 + \alpha_2 + \alpha_3}, \tag{3.39}$$

$$\omega_2 = \frac{\alpha_2}{\alpha_1 + \alpha_2 + \alpha_3}, \tag{3.40}$$

and

$$\omega_3 = \frac{\alpha_3}{\alpha_1 + \alpha_2 + \alpha_3} \tag{3.41}$$

gives (approximately) the optimal weights of $\omega_1 = 0.1$, $\omega_2 = 0.6$ and $\omega_3 = 0.3$ when the S_k are small enough to be dominated by ϵ. Nearly optimal weights are also obtained when the S_k are larger than ϵ, as long as all the S_k are approximately the same size, as is the case for sufficiently smooth data. On the other hand, if the data are not smooth as indicated by large S_k, then the corresponding α_k will be small compared to the other α_k's, giving that particular stencil limited influence. If two of the S_k are relatively large, then their corresponding α_k's will both be small, and the scheme will rely most heavily on a single stencil similar to the digital behavior of HJ ENO. In the unfortunate instance that all three of the S_k are large, the data are poorly conditioned, and none of the stencils are particularly useful. This case is problematic for the HJ ENO method as well, but fortunately it usually occurs only locally in space and time, allowing the methods to repair themselves after the situation subsides.

The function $(\phi_x^+)_i$ is constructed with a subset of $\{\phi_{i-2}, \phi_{i-1}, \phi_i, \phi_{i+1}, \phi_{i+2}, \phi_{i+3}\}$. Defining $v_1 = D^+\phi_{i+2}$, $v_2 = D^+\phi_{i+1}$, $v_3 = D^+\phi_i$, $v_4 = D^+\phi_{i-1}$, and $v_5 = D^+\phi_{i-2}$ allows us to use equations (3.25), (3.26), and (3.27) as the three HJ ENO approximations to $(\phi_x^+)_i$. Then the HJ WENO convex combination is given by equation (3.28) with the weights given by equations (3.39), (3.40), and (3.41).

3.5 TVD Runge-Kutta

HJ ENO and HJ WENO allow us to discretize the spatial terms in equation (3.2) to fifth-order accuracy, while the forward Euler time discretization in equation (3.4) is only first-order accurate in time. Practical experience suggests that level set methods are sensitive to spatial accuracy, implying that the fifth-order accurate HJ WENO method is desirable. On the other hand, temporal truncation errors seem to produce significantly less deterioration of the numerical solution, so one can often use the low-order accurate forward Euler method for discretization in time.

There are times when a higher-order temporal discretization is necessary in order to obtain accurate numerical solutions. In [150], Shu and Osher proposed total variation diminishing (TVD) Runge-Kutta (RK) methods to increase the accuracy for a *method of lines* approach to temporal discretization. The method of lines approach assumes that the spatial discretization can be separated from the temporal discretization in a semidiscrete manner that allows the temporal discretization of the PDE to be treated independently as an ODE. While there are numerous RK schemes, these TVD RK schemes guarantee that no spurious oscillations are produced as a consequence of the higher-order accurate temporal discretization as long as no spurious oscillations are produced with the forward Euler building block.

The basic first-order accurate TVD RK scheme is just the forward Euler method. As mentioned above, we assume that the forward Euler method is TVD in conjunction with the spatial discretization of the PDE. Then higher-order accurate methods are obtained by sequentially taking Euler steps and combining the results with the initial data using a *convex combination*. Since the Euler steps are TVD (by assumption) and the convex combination operation is TVD as long as the coefficients are positive, the resulting higher-order accurate TVD RK method is TVD. Unfortunately, in our specific case, the HJ ENO and HJ WENO schemes are not TVD when used in conjunction with upwinding to approximate equation (3.4). However, practical numerical experience has shown that the HJ ENO and HJ WENO schemes are most likely total variation bounded (TVB), implying that the overall method is also TVB using the TVD RK schemes.

The second-order accurate TVD RK scheme is identical to the standard second-order accurate RK scheme. It is also known as the midpoint rule, as the modified Euler method, and as Heun's predictor-corrector method. First, an Euler step is taken to advance the solution to time $t^n + \triangle t$,

$$\frac{\phi^{n+1} - \phi^n}{\triangle t} + \vec{V}^n \cdot \nabla \phi^n = 0, \tag{3.42}$$

followed by a second Euler step to advance the solution to time $t^n + 2\triangle t$,

$$\frac{\phi^{n+2} - \phi^{n+1}}{\triangle t} + \vec{V}^{n+1} \cdot \nabla \phi^{n+1} = 0, \tag{3.43}$$

followed by an averaging step

$$\phi^{n+1} = \frac{1}{2}\phi^n + \frac{1}{2}\phi^{n+2} \tag{3.44}$$

that takes a convex combination of the initial data and the result of two Euler steps. The final averaging step produces the second-order accurate TVD (or TVB for HJ ENO and HJ WENO) approximation to ϕ at time $t^n + \triangle t$.

The third-order accurate TVD RK scheme proposed in [150] is as follows. First, an Euler step is taken to advance the solution to time $t^n + \triangle t$,

$$\frac{\phi^{n+1} - \phi^n}{\triangle t} + \vec{V}^n \cdot \nabla \phi^n = 0, \tag{3.45}$$

followed by a second Euler step to advance the solution to time $t^n + 2\triangle t$,

$$\frac{\phi^{n+2} - \phi^{n+1}}{\triangle t} + \vec{V}^{n+1} \cdot \nabla \phi^{n+1} = 0, \tag{3.46}$$

followed by an averaging step

$$\phi^{n+\frac{1}{2}} = \frac{3}{4}\phi^n + \frac{1}{4}\phi^{n+2} \tag{3.47}$$

that produces an approximation to ϕ at time $t^n + \frac{1}{2}\triangle t$. Then another Euler step is taken to advance the solution to time $t^n + \frac{3}{2}\triangle t$,

$$\frac{\phi^{n+\frac{3}{2}} - \phi^{n+\frac{1}{2}}}{\triangle t} + \vec{V}^{n+\frac{1}{2}} \cdot \nabla \phi^{n+\frac{1}{2}} = 0, \tag{3.48}$$

followed by a second averaging step

$$\phi^{n+1} = \frac{1}{3}\phi^n + \frac{2}{3}\phi^{n+\frac{3}{2}} \tag{3.49}$$

that produces a third-order accurate approximation to ϕ at time $t^n + \triangle t$. This third-order accurate TVD RK method has a stability region that includes part of the imaginary axis. Thus, a stable (although ill-advised) numerical method results from combining third-order accurate TVD RK with central differencing for the spatial discretization.

While fourth-order accurate (and higher) TVD RK schemes exist, this improved temporal accuracy does not seem to make a significant difference in practical calculations, especially since the HJ WENO scheme usually loses accuracy and looks a lot like the third-order accurate HJ ENO scheme in many interesting areas of the flow. Also, the fourth-order accurate (and higher) TVD RK methods require both upwind and downwind differencing approximations, doubling the computational cost of evaluating the spatial operators. See [150] for fourth- and fifth-order accurate TVD RK schemes. Finally, we note that a rather interesting approach to TVD RK schemes has recently been carried out by Spiteri and Ruuth [154], who proposed increasing the number of internal stages so that this number exceeds the order of the method.

4

Motion Involving Mean Curvature

4.1 Equation of Motion

In the last chapter we discussed the motion of an interface in an externally generated velocity field $\vec{V}(\vec{x}, t)$. In this chapter we discuss interface motion for a self-generated velocity field $\vec{V}$ that depends directly on the level set function ϕ. As an example, we consider motion by mean curvature where the interface moves in the normal direction with a velocity proportional to its curvature; i.e., $\vec{V} = -b\kappa\vec{N}$, where $b > 0$ is a constant and κ is the curvature. When $b > 0$, the interface moves in the direction of concavity, so that circles (in two dimensions) shrink to a single point and disappear. When $b < 0$, the interface moves in the direction of convexity, so that circles grow instead of shrink. This growing-circle effect leads to the growth of small perturbations in the front including those due to round-off errors. Because $b < 0$ allows small erroneous perturbations to incorrectly grow into $O(1)$ features, the $b < 0$ case is *ill-posed*, and we do not consider it here. Figure 4.1 shows the motion of a wound spiral in a curvature-driven flow. The high-curvature ends of the spiral move significantly faster than the relatively low curvature elongated body section. Figure 4.2 shows the evolution of a star-shaped interface in a curvature-driven flow. The tips of the star move inward, while the gaps in between the tips move outward.

The velocity field for motion by mean curvature contains a component in the normal direction only, i.e., the tangential component is identically zero. In general, one does not need to specify tangential components when devising a velocity field. Since $\vec{N}$ and $\nabla\phi$ point in the same direction,

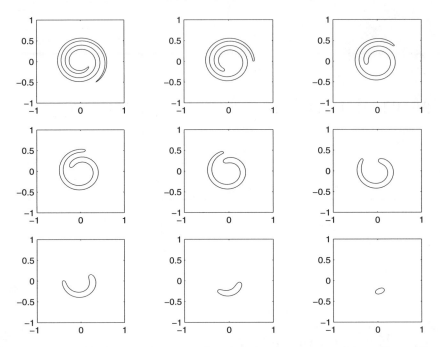

Figure 4.1. Evolution of a wound spiral in a curvature-driven flow. The high-curvature ends of the spiral move significantly faster than the elongated body section.

$\vec{T} \cdot \nabla\phi = 0$ for any tangent vector $\vec{T}$, implying that the tangential velocity components vanish when plugged into the level set equation. For example, in two spatial dimensions with $\vec{V} = V_n\vec{N} + V_t\vec{T}$, the level set equation

$$\phi_t + \left(V_n\vec{N} + V_t\vec{T}\right) \cdot \nabla\phi = 0 \tag{4.1}$$

is equivalent to

$$\phi_t + V_n\vec{N} \cdot \nabla\phi = 0, \tag{4.2}$$

since $\vec{T} \cdot \nabla\phi = 0$. Furthermore, since

$$\vec{N} \cdot \nabla\phi = \frac{\nabla\phi}{|\nabla\phi|} \cdot \nabla\phi = \frac{|\nabla\phi|^2}{|\nabla\phi|} = |\nabla\phi|, \tag{4.3}$$

we can rewrite equation (4.2) as

$$\phi_t + V_n|\nabla\phi| = 0 \tag{4.4}$$

where V_n is the component of velocity in the normal direction, otherwise known as the *normal velocity*. Thus, motion by mean curvature is characterized by $V_n = -b\kappa$.

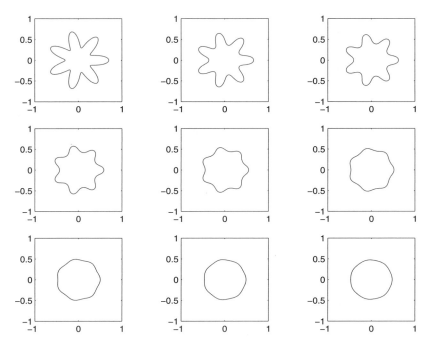

Figure 4.2. Evolution of a star-shaped interface in a curvature-driven flow. The tips of the star move inward, while the gaps in between the tips move outward.

Equation (4.4) is also known as the equation of the level set equation. Like equation (3.2), equation (3.2) is used for externally generated velocity fields, while equation (4.4) is used for (internally) self-generated velocity fields. As we shall see shortly, this is more than a notational difference. In fact, slightly more complicated numerical methods are needed for equation (4.4) than were proposed in the last chapter for equation (3.2).

Plugging $V_n = -b\kappa$ into the level set equation (4.4) gives

$$\phi_t = b\kappa|\nabla\phi|, \tag{4.5}$$

where we have moved the spatial term to the right-hand side. We note that $b\kappa|\nabla\phi|$ is a *parabolic* term that cannot be discretized with an upwind approach. When ϕ is a signed distance function, equation (4.5) becomes the heat equation

$$\phi_t = b\triangle\phi, \tag{4.6}$$

where ϕ is the temperature and b is the thermal conductivity. The heat equation is the most basic equation of the parabolic model.

When ϕ is a signed distance function, $b\kappa|\nabla\phi|$ and $b\triangle\phi$ are identical, and either of these can be used to calculate the right-hand side of equation (4.5). However, once this right-hand side is combined with a forward Euler time step (or a forward Euler substep in the case of RK), the new value of ϕ is not a signed distance function, and equations (4.5) and (4.6) can no longer be interchanged. If this new value of ϕ is reinitialized to a signed distance function (methods for doing this are outlined in Chapter 7), then $b\triangle\phi$ can be used in place of $b\kappa|\nabla\phi|$ in the next time step as well. In summary, equations (4.5) and (4.6) have the *same* effect on the interface location as long as one keeps ϕ equal to the signed distance function off the interface. Note that keeping ϕ equal to signed distance off the interface does not change the interface location. It only changes the implicit embedding function used to identify the interface location.

4.2   Numerical Discretization

Parabolic equations such as the heat equation need to be discretized using central differencing since the domain of dependence includes information from all spatial directions, as opposed to hyperbolic equations like equation (3.2), where information flows in the direction of characteristics only. Thus, the $\triangle\phi$ term in equation (4.6) is discretized using the second-order accurate formula in equation (1.9) in each spatial dimension (see equation (2.7)). A similar approach should therefore be taken in discretizing equation (4.5). The curvature κ is discretized using second-order accurate central differencing as outlined in equation (1.8) and the discussion following that equation. Likewise, the $\nabla\phi$ term is discretized using the second order accurate central differencing in equation (1.5) applied independently in each spatial dimension. While these discretizations are only second-order accurate in space, the dissipative nature of the equations usually makes this second-order accuracy sufficient.

Central differencing of $\triangle\phi$ in equation (4.6) combined with a forward Euler time discretization requires a time-step restriction of

$$\triangle t\left(\frac{2b}{(\triangle x)^2} + \frac{2b}{(\triangle y)^2} + \frac{2b}{(\triangle z)^2}\right) < 1 \tag{4.7}$$

to maintain stability of the numerical algorithm. Here $\triangle t$ is $O((\triangle x)^2)$, which is significantly more stringent than in the hyperbolic case, where $\triangle t$ is only $O(\triangle x)$. Enforcing $\triangle t = O((\triangle x)^2)$ gives an overall $O((\triangle x)^2)$ accurate discretization, even though forward Euler is used for the time differencing (i.e., since the first-order accurate $O(\triangle t)$ time discretization is $O((\triangle x)^2)$). Equation (4.5) can be discretized using forward Euler time stepping with the CFL condition in equation (4.7) as well.

The stringent $O((\triangle x)^2)$ time-step restriction resulting from the forward Euler time discretization can be alleviated by using an ODE solver with a larger stability region, e.g., an implicit method. For example, first-order accurate *backward Euler* time stepping applied to equation (4.6) gives

$$\frac{\phi^{n+1} - \phi^n}{\Delta t} = b\triangle\phi^{n+1}, \tag{4.8}$$

which has no time step stability restriction on the size of $\triangle t$. This means that $\triangle t$ can be chosen for accuracy reasons alone, and one typically sets $\triangle t = O(\triangle x)$. Note that setting $\triangle t = O(\triangle x)$ as opposed to $\triangle t = O((\triangle x)^2)$ lowers the overall accuracy to $O(\triangle x)$. This can be improved upon using the *trapezoidal rule*

$$\frac{\phi^{n+1} - \phi^n}{\Delta t} = b\left(\frac{\triangle\phi^n + \triangle\phi^{n+1}}{2}\right), \tag{4.9}$$

which is $O((\triangle t)^2)$ in time and thus $O((\triangle x)^2)$ overall even when $\triangle t = O(\triangle x)$. This combination of the trapezoidal rule with central differencing of a parabolic spatial operator is generally referred to as the *Crank-Nicolson* scheme.

The price we pay for the larger time step achieved using either equation (4.8) or equation (4.9) is that a linear system of equations must be solved at each time step to obtain ϕ^{n+1}. Luckily, this is not difficult given the simple linear structure of $\triangle\phi^{n+1}$. Unfortunately, an implicit discretization of equation (4.5) requires consideration of the more complicated nonlinear $\kappa^{n+1}|\nabla\phi^{n+1}|$ term.

We caution the reader that one cannot substitute equation (4.6) for equation (4.5) when using an implicit time discretization. Even if ϕ^n is initially a signed distance function, ϕ^{n+1} will generally not be a signed distance function after the linear system has been solved. This means that $\triangle\phi^{n+1}$ is not a good approximation to $\kappa^{n+1}|\nabla\phi^{n+1}|$ even though $\triangle\phi^n$ may be exactly equal to $\kappa^n|\nabla\phi^n|$. Although we stress (throughout the book) the conceptual simplifications and computational savings that can be obtained when ϕ is a signed distance function, e.g., replacing $\vec{N}$ with $\nabla\phi$, κ with $\triangle\phi$, etc., we caution the reader that there is a significant and important difference between the two in the case where ϕ is not a signed distance function.

4.3   Convection-Diffusion Equations

The *convection-diffusion equation*

$$\phi_t + \vec{V} \cdot \nabla\phi = b\triangle\phi \tag{4.10}$$

includes both the effects of an external velocity field and a diffusive term. The level set version of this is

$$\phi_t + \vec{V} \cdot \nabla\phi = b\kappa|\nabla\phi|, \tag{4.11}$$

and the two can be used interchangeably if one maintains a signed distance approximation for ϕ off the interface. These equations can be solved using the upwind methods from the last chapter on the $\vec{V} \cdot \nabla \phi$ term and central differencing on the parabolic $b \triangle \phi$ or $b \kappa |\nabla \phi|$ term. A TVD RK time discretization can be used with a time-step restriction of

$$\triangle t \left(\frac{|u|}{\triangle x} + \frac{|v|}{\triangle y} + \frac{|w|}{\triangle z} + \frac{2b}{(\triangle x)^2} + \frac{2b}{(\triangle y)^2} + \frac{2b}{(\triangle z)^2} \right) < 1 \qquad (4.12)$$

satisfied at every grid point.

Suppose the $O(1)$ size b term is replaced with an $O(\triangle x)$ size ϵ term that vanishes as the mesh is refined with $\triangle x \to 0$. Then equation (4.10) becomes

$$\phi_t + \vec{V} \cdot \nabla \phi = \epsilon \triangle \phi, \qquad (4.13)$$

which asymptotically approaches equation (3.2) as $\epsilon \to 0$. The addition of an artificial $\epsilon \triangle \phi$ term to the right-hand side of equation (3.2) is called the *artificial viscosity* method. Artificial viscosity is used by many authors to stabilize a central differencing approximation to the convective $\nabla \phi$ term in equation (3.2). This arises in computational fluid dynamics, where terms of the form $\epsilon \triangle \phi$ are added to the right-hand side of convective equations to pick out *vanishing viscosity solutions* valid in the limit as $\epsilon \to 0$. This vanishing viscosity picks out the physically correct *weak solution* when no classical solution exists, for example in the case of a discontinuous shock wave. It is interesting to note that the upwind discretizations discussed in the last chapter have numerical truncation errors that serve the same purpose as the $\epsilon \triangle \phi$ term. First-order accurate upwinding has an intrinsic $O(\triangle x)$ artificial viscosity, and the higher-order accurate upwind methods have intrinsic artificial viscosities with magnitude $O((\triangle x)^r)$, where r is the order of accuracy of the method.

In [146], Sethian suggested an entropy condition that required curves to flow into corners, and he provided numerical evidence to show that this entropy condition produced the correct weak solution for self-interesting curves. Sethian's entropy condition indicates that $\epsilon \kappa |\nabla \phi|$ is a better form for the vanishing viscosity than $\epsilon \triangle \phi$ for dealing with the evolution of lower-dimensional interfaces. This concept was rigorized by Osher and Sethian in [126], where they pointed out that

$$\phi_t + \vec{V} \cdot \nabla \phi = \epsilon \kappa |\nabla \phi| \qquad (4.14)$$

is a more natural choice than equation (4.13) for dealing with level set methods, although these two equations are interchangeable when ϕ is a signed distance function.

5

Hamilton-Jacobi Equations

5.1 Introduction

In this chapter we discuss numerical methods for the solution of general *Hamilton-Jacobi* equations of the form

$$\phi_t + H(\nabla\phi) = 0, \tag{5.1}$$

where H can be a function of both space and time. In three spatial dimensions, we can write

$$\phi_t + H(\phi_x, \phi_y, \phi_z) = 0 \tag{5.2}$$

as an expanded version of equation (5.1). Convection in an externally generated velocity field (equation (3.2)) is an example of a Hamilton-Jacobi equation where $H(\nabla\phi) = \vec{V} \cdot \nabla\phi$. The level set equation (equation (4.4)) is another example of a Hamilton-Jacobi equation with $H(\nabla\phi) = V_n|\nabla\phi|$. Here V_n can depend on $\vec{x}$, t, or even $\nabla\phi/|\nabla\phi|$.

The equation for motion by mean curvature (equation (4.5)) is not a Hamilton-Jacobi-type equation, since the front speed depends on the second derivatives of ϕ. Hamilton-Jacobi equations depend on (at most) the first derivatives of ϕ, and these equations are hyperbolic. The equation for motion by mean curvature depends on the second derivatives of ϕ and is parabolic.

5.2   Connection with Conservation Laws

Consider the one-dimensional scalar *conservation law*

$$u_t + F(u)_x = 0; \tag{5.3}$$

where u is the *conserved quantity* and $F(u)$ is the *flux* function. A well-known conservation law is the *continuity equation*

$$\rho_t + (\rho u)_x = 0 \tag{5.4}$$

for *conservation of mass*, where ρ is the density of the material. In *computational fluid dynamics* (*CFD*), the continuity equation is combined with equations for *conservation of momentum* and *conservation of energy* to obtain the compressible *Navier-Stokes* equations. When viscous effects are ignored, the Navier-Stokes equations reduce to the compressible inviscid *Euler* equations.

The presence of discontinuities in the Euler equations forces one to consider weak solutions where the derivatives of solution variables, e.g., ρ_x, can fail to exist. Examples include linear *contact* discontinuities and nonlinear *shock* waves. The nonlinear nature of shock waves allows them to develop as the solution progresses forward in time even if the data are initially smooth. The Euler equations may not always have unique solutions, and an *entropy condition* is used to pick out the physically correct solution. This is the vanishing viscosity solution discussed in the last chapter. For example, vanishing viscosity admits physically consistent *rarefaction* waves, ruling out physically inadmissible *expansion shocks*.

Burgers' equation

$$u_t + \left(\frac{u^2}{2}\right)_x = 0 \tag{5.5}$$

is a scalar conservation law that possesses many of the interesting nonlinear properties contained in the more complex Euler equations. Burgers' equation develops discontinuous shock waves from smooth initial data and exhibits nonphysical expansion shocks if the vanishing viscosity solution is not used to force these to become smooth rarefaction waves. Many of the numerical methods developed to solve Burgers' equation can be extended to treat both the one-dimensional and the multidimensional Euler equations of gas dynamics.

Consider the one-dimensional Hamilton-Jacobi equation

$$\phi_t + H(\phi_x) = 0, \tag{5.6}$$

which becomes

$$(\phi_x)_t + H(\phi_x)_x = 0 \tag{5.7}$$

after one takes a spatial derivative of the entire equation. Setting $u = \phi_x$ in equation 5.7 results in

$$u_t + H(u)_x = 0, \tag{5.8}$$

which is a scalar conservation law; see equation (5.3). Thus, in one spatial dimension we can draw a direct correspondence between Hamilton-Jacobi equations and conservation laws. The solution u to a conservation law is the derivative of a solution ϕ to a Hamilton-Jacobi equation. Conversely, the solution ϕ to a Hamilton-Jacobi equation is the integral of a solution u to a conservation law. This allows us to point out a number of useful facts. For example, since the integral of a discontinuity is a *kink*, or discontinuity in the first derivative, solutions to Hamilton-Jacobi equations can develop kinks in the solution even if the data are initially smooth. In addition, solutions to Hamilton-Jacobi equations cannot generally develop a discontinuity unless the corresponding conservation law develops a delta function. Thus, solutions ϕ to equation (5.2) are typically continuous. Furthermore, since conservation laws can have nonunique solutions, entropy conditions are needed to pick out "physically" relevant solutions to equation (5.2) as well.

Viscosity solutions for Hamilton-Jacobi equations were first proposed by Crandall and Lions [52]. Monotone first-order accurate numerical methods were first presented by Crandall and Lions [53] as well. Later, Osher and Sethian [126] used the connection between conservation laws and Hamilton-Jacobi equations to construct higher-order accurate "artifact-free" numerical methods. Even though the analogy between conservation laws and Hamilton-Jacobi equations fails in multiple spatial dimensions, many Hamilton-Jacobi equations can be discretized in a dimension by dimension fashion. This culminated in [127], where Osher and Shu proposed a general framework for the numerical solution of Hamilton-Jacobi equations using successful methods from the theory of conservation laws. We follow [127] below.

5.3 Numerical Discretization

A forward Euler time discretization of a Hamilton-Jacobi equation can be written as

$$\frac{\phi^{n+1} - \phi^n}{\triangle t} + \hat{H}^n(\phi_x^-, \phi_x^+; \phi_y^-, \phi_y^+; \phi_z^-, \phi_z^+) = 0, \tag{5.9}$$

where $\hat{H}(\phi_x^-, \phi_x^+; \phi_y^-, \phi_y^+; \phi_z^-, \phi_z^+)$ is a numerical approximation of $H(\phi_x, \phi_y, \phi_z)$. The function $\hat{H}$ is called a numerical Hamiltonian, and it is required to be consistent in the sense that $\hat{H}(\phi_x, \phi_x; \phi_y, \phi_y; \phi_z, \phi_z) = H(\phi_x, \phi_y, \phi_z)$. Recall that spatial derivatives such as ϕ_x^- are discretized with either first-order accurate one-sided differencing or the higher-order accurate HJ ENO or HJ WENO schemes. For brevity, we discuss the two-dimensional numerical approximation to $H(\phi_x, \phi_y)$, noting that the extension to three spatial dimensions is straightforward. An important class of schemes is that of *monotone schemes*. A scheme is monotone when ϕ^{n+1} as defined in equation (5.9) is a nondecreasing function of all the ϕ^n. Crandall and Lions proved that these schemes converge to the correct solution, although they are only first-order accurate. The numerical Hamiltonians associated with monotone schemes are important, and examples will be given below.

The forward Euler time discretization (equation (5.9)) can be extended to higher-order TVD Runge Kutta in a straightforward manner, as discussed in Chapter (3). The CFL condition for equation 5.9 is

$$\Delta t \max \left\{ \frac{|H_1|}{\Delta x} + \frac{|H_2|}{\Delta y} + \frac{|H_3|}{\Delta z} \right\} < 1, \tag{5.10}$$

where H_1, H_2, and H_3 are the partial derivatives of H with respect to ϕ_x, ϕ_y, and ϕ_z, respectively. For example, in equation (3.2), where $H(\nabla \phi) = \vec{V} \cdot \nabla \phi$, the partial derivatives of H are $H_1 = u$, $H_2 = v$, and $H_3 = w$. In this case equation (5.10) reduces to equation (3.10). As another example, consider the level set equation (4.4) with $H(\nabla \phi) = V_n |\nabla \phi|$. Here the partial derivatives are slightly more complicated, with $H_1 = V_N \phi_x / |\nabla \phi|$, $H_2 = V_N \phi_y / |\nabla \phi|$, and $H_3 = V_N \phi_z / |\nabla \phi|$, assuming that V_N does not depend on ϕ_x, ϕ_y or ϕ_z. Otherwise, the partial derivatives can be substantially more complicated.

5.3.1   Lax-Friedrichs Schemes

The first approximation to $\hat{H}$ that we consider is the *Lax-Friedrichs* (LF) scheme from [53] given by

$$\hat{H} = H \left(\frac{\phi_x^- + \phi_x^+}{2}, \frac{\phi_y^- + \phi_y^+}{2} \right) - \alpha^x \left(\frac{\phi_x^+ - \phi_x^-}{2} \right) - \alpha^y \left(\frac{\phi_y^+ - \phi_y^-}{2} \right), \tag{5.11}$$

where α^x and α^y are dissipation coefficients that control the amount of numerical viscosity. These dissipation coefficients

$$\alpha^x = \max |H_1(\phi_x, \phi_y)|, \qquad \alpha^y = \max |H_2(\phi_x, \phi_y)| \tag{5.12}$$

are chosen based on the partial derivatives of H.

The choice of the dissipation coefficients in equation (5.12) can be rather subtle. In the traditional implementation of the LF scheme, the maximum is chosen over the entire computational domain. First, the maximum and minimum values of ϕ_x are identified by considering all the values of ϕ_x^- and ϕ_x^+ on the Cartesian mesh. Then one can identify the interval $I^x = [\phi_x^{\min}, \phi_x^{\max}]$. A similar procedure is used to define $I^y = [\phi_y^{\min}, \phi_y^{\max}]$. The coefficients α^x and α^y are set to the maximum possible values of $|H_1(\phi_x, \phi_y)|$ and $|H_2(\phi_x, \phi_y)|$, respectively, with $\phi_x \in I^x$ and $\phi_y \in I^y$. Although it is occasionally difficult to evaluate the maximum values of $|H_1|$ and $|H_2|$, it is straightforward to do so in many instances. For example, in equation (3.2), both $H_1 = u$ and $H_2 = v$ are independent of ϕ_x and ϕ_y, so α^x and α^y can be set to the maximum values of $|u|$ and $|v|$ on the Cartesian mesh.

Consider evaluating α^x and α^y for equation (4.4) where $H_1 = V_N\phi_x/|\nabla\phi|$ and $H_2 = V_N\phi_y/|\nabla\phi|$, recalling that these are the partial derivatives only if V_N is independent ϕ_x and ϕ_y. It is somewhat more complicated to evaluate α^x and α^y in this case. When ϕ is a signed distance function with $|\nabla\phi| = 1$ (or ≈ 1 numerically), we can simplify to $H_1 = V_N\phi_x$ and $H_2 = V_N\phi_y$. These functions can still be somewhat tricky to work with if V_N is spatially varying. But in the special case that V_N is spatially constant, the maximum values of $|H_1|$ and $|H_2|$ can be determined by considering only the endpoints of I_x and I_y, respectively. This is true because H_1 and H_2 are monotone functions of ϕ_x and ϕ_y, respectively. In fact, when V_N is spatially constant, $H_1 = V_N\phi_x/|\nabla\phi|$ and $H_2 = V_N\phi_y/|\nabla\phi|$ are straightforward to work with as well. The function H_1 achieves a maximum when $|\phi_x|$ is as large as possible and $|\phi_y|$ is as small as possible. Thus, only the endpoints of I^x and I^y need be considered; note that we use $\phi_y = 0$ when the endpoints of I^y differ in sign. Similar reasoning can be used to find the maximum value of $|H_2|$. One way to treat a spatially varying V_N is to make some estimates. For example, since $|\phi_x|/|\nabla\phi| \leq 1$ for all ϕ_x and ϕ_y, we can bound $|H_1| \leq |V_N|$. A similar bound of $|H_2| \leq |V_N|$ holds for $|H_2|$. Thus, both α^x and α^y can be set to the maximum value of $|V_N|$ on the Cartesian mesh. The price we pay for using bounds to choose α larger than it should be is increased numerical dissipation. That is, while the numerical method will be stable and give an accurate solution as the mesh is refined, some details of this solution may be smeared out and lost on a coarser mesh.

Since increasing α increases the amount of artificial dissipation, decreasing the quality of the solution, it is beneficial to chose α as small as possible without inducing oscillations or other nonphysical phenomena into the solution. In approximating $\hat{H}_{i,j}$ at a grid point $\vec{x}_{i,j}$ on a Cartesian mesh, it then makes little sense to do a global search to define the intervals I^x and I^y. In particular, consider the simple convection equation (3.2) where $\alpha^x = \max|u|$ and $\alpha^y = \max|v|$. Suppose that some region had relatively small values of $|u|$ and $|v|$, while another region had relatively large values. Since the LF method chooses α^x as the largest value of $|u|$ and α^y as the largest value of $|v|$, the same values of α will be used in the region where the velocities are small as is used in the region where the velocities are large. In the region where the velocities are large, the large values of α are required to obtain a good solution. But in the region where the velocities are small, these large values of α produce too much numerical dissipation, wiping out small features of the solution. Thus, it is advantageous to use only the grid points sufficiently close to $\vec{x}_{i,j}$ in determining α. A rule of thumb is to include the grid points from $\vec{x}_{i-3,j}$ to $\vec{x}_{i+3,j}$ in the x-direction and from $\vec{x}_{i,j-3}$ to $\vec{x}_{i,j+3}$ in the y-direction in the local search neighborhood for determining α. This includes all the grid nodes that are used to evaluate

$\phi_x^{\pm}$ and $\phi_y^{\pm}$ at $\vec{x}_{i,j}$ using the HJ WENO scheme. This type of scheme has been referred to as a *Stencil Lax-Friedrichs* (SLF) scheme, since it determines the dissipation coefficient using only the neighboring grid points that are part of the stencil used to determine ϕ_x and ϕ_y. An alternative to the dimension-by-dimension neighborhoods is to use the 49 grid points in the rectangle with diagonal corners at $\vec{x}_{i-3,j-3}$ and $\vec{x}_{i+3,j+3}$ to determine α.

This idea of searching only locally to determine the dissipation coefficients can be taken a step further. The *Local Lax-Friedrichs* (LLF) scheme proposed for conservation laws by Shu and Osher [151] does not look at any neighboring grid points when calculating the dissipation coefficients in a given direction. In [127], Osher and Shu interpreted this to mean that α^x is determined at each grid point using only the values of ϕ_x^- and ϕ_x^+ at that specific grid point to determine the interval I^x. The interval I^y is still determined in the LF or SLF manner (in the SLF case we rename LLF as SLLF). Similarly, α^y uses an interval I^y, defined using only the values of ϕ_y^- and ϕ_y^+ at the grid point in question while I^x is still determined in the LF or SLF fashion. Osher and Shu [127] also proposed the *Local Local Lax-Friedrichs* (LLLF) scheme with even less numerical dissipation. At each grid point I^x is determined using the values of ϕ_x^- and ϕ_x^+ at that grid point; I^y is determined using the values of ϕ_y^- and ϕ_y^+ at that grid point; and then these intervals are used to determine both α^x and α^y. When H is separable, i.e., $H(\phi_x, \phi_y) = H^x(\phi_x) + H^y(\phi_y)$, LLLF reduces to LLF, since α^x is independent of ϕ_y, and α^y is independent of ϕ_x. When H is not separable, LLF and LLLF are truly distinct schemes. In practice, LLF seems to work better than any of the other options. LF and SLF are usually too dissipative, while LLLF is usually not dissipative enough to overcome the problems introduced by using the centrally averaged approximation to ϕ_x and ϕ_y in evaluating H in equation (5.11). Note that LLF is a monotone scheme.

5.3.2   The Roe-Fix Scheme

As discussed above, choosing the appropriate amount of artificial dissipation to add to the centrally evaluated H in equation (5.11) can be tricky. Therefore, it is often desirable to use upwind-based methods with built-in artificial dissipation. For conservation laws, Shu and Osher [151] proposed using Roe's upwind method along with an LLF entropy correction at sonic points where entropy-violating expansion shocks might form. The added dissipation from the LLF entropy correction forces the expansion shocks to develop into continuous rarefaction waves. The method was dubbed *Roe Fix* (RF) and it can be written for Hamilton-Jacobi equations (see [127]) as

$$\hat{H} = H\left(\phi_x^{\star}, \phi_y^{\star}\right) - \alpha^x \left(\frac{\phi_x^+ - \phi_x^-}{2}\right) - \alpha^y \left(\frac{\phi_y^+ - \phi_y^-}{2}\right), \qquad (5.13)$$

where α^x and α^y are usually set identically to zero in order to remove the numerical dissipation terms. In the RF scheme, I^x and I^y are initially determined using only the nodal values for $\phi_x^{\pm}$ and $\phi_y^{\pm}$ as in the LLLF scheme. In order to estimate the potential for upwinding, we look at the partial derivatives H_1 and H_2. If $H_1(\phi_x, \phi_y)$ has the same sign (either always positive or always negative) for all $\phi_x \in I^x$ and all $\phi_y \in I^y$, we know which way information is flowing and can apply upwinding. Similarly, if $H_2(\phi_x, \phi_y)$ has the same sign for all $\phi_x \in I^x$ and $\phi_y \in I^y$, we can upwind this term as well. If both H_1 and H_2 do not change sign, we upwind completely, setting both α^x and α^y to zero. If $H_1 > 0$, information is flowing from left to right, and we set $\phi_x^{\star} = \phi_x^{-}$. Otherwise, $H_1 < 0$, and we set $\phi_x^{\star} = \phi_x^{+}$. Similarly, $H_2 > 0$ indicates $\phi_y^{\star} = \phi_y^{-}$, and $H_2 < 0$ indicates $\phi_y^{\star} = \phi_y^{+}$.

If either H_1 or H_2 changes sign, we are in the vicinity of a *sonic point* where the *eigenvalue* (in this case H_1 or H_2) is identically zero. This signifies a potential difficulty with nonunique solutions, and artificial dissipation is needed to pick out the physically correct vanishing viscosity solution. We switch from the RF scheme to the LLF scheme to obtain the needed artificial viscosity. If there is a sonic point in only one direction, i.e., x or y, it makes little sense to add damping in both directions. Therefore, we look for sonic points in each direction and add damping only to the directions that have sonic points. This is done using the I^x and I^y defined as in the LLF method. That is, we switch from the LLLF defined intervals used above to initially look for sonic points to the larger LLF intervals that are even more likely to have sonic points. We proceed as follows. If $H_1(\phi_x, \phi_y)$ does not change sign for all $\phi_x \in I_{LLF}^x$ and all $\phi_y \in I_{LLF}^y$, we set $\phi_x^{\star}$ equal to either ϕ_x^{-} or ϕ_x^{+} depending on the sign of H_1. In addition, we set α^x to zero to remove the artificial dissipation in the x-direction. At the same time, this means that a sonic point must have occurred in H_2, so we use an LLF-type method for the y-direction, setting $\phi_y^{\star} = (\phi_y^{-} + \phi_y^{+})/2$ and choosing α^y as dictated by the LLF scheme. A similar algorithm is executed if $H_2(\phi_x, \phi_y)$ does not change sign for all $\phi_x \in I_{LLF}^x$ and $\phi_y \in I_{LLF}^y$. Then $\phi_y^{\star}$ is set to either ϕ_y^{-} or ϕ_y^{+}, depending on the sign of H_2; α^y is set to zero; and an LLF method is used in the x-direction, setting $\phi_x^{\star} = (\phi_x^{-} + \phi_x^{+})/2$ while choosing α^x as dictated by the LLF scheme. If both H_1 and H_2 change sign, we have sonic points in both directions and proceed with the standard LLF scheme at that grid point.

With the RF scheme, upwinding in the x-direction dictates that either ϕ_x^{-} or ϕ_x^{+} be used, but not both. Similarly, upwinding in the y-direction uses either ϕ_y^{-} or ϕ_y^{+}, but not both. Since evaluating $\phi_x^{\pm}$ and $\phi_y^{\pm}$ using high-order accurate HJ ENO or HJ WENO schemes is rather costly, it seems wasteful to do twice as much work in these instances. Unfortunately, one cannot determine whether upwinding can be used (as opposed to LLF) without computing $\phi_x^{\pm}$ and $\phi_y^{\pm}$. In order to minimize CPU time, one can compute $\phi_x^{\pm}$ and $\phi_y^{\pm}$ using the first-order accurate forward and backward difference formulas and use these cheaper approximations to decide whether or not upwinding or LLF will be used. After making this decision, the higher-order accurate HJ WENO (or HJ ENO) method can be used to compute the necessary values of $\phi_x^\pm$ and $\phi_y^\pm$ used in the numerical discretization, obtaining the usual high-order accuracy. Sonic points rarely occur in practice, and this strategy reduces the use of the costly HJ WENO method by approximately a factor of two.

5.3.3 Godunov's Scheme

In [74], Godunov proposed a numerical method that gives the exact solution to the *Riemann problem* for one-dimensional conservation laws with piecewise constant initial data. The multidimensional Hamilton-Jacobi formulation of this scheme can be written as

$$\hat{H} = \text{ext}_x \text{ext}_y H \left(\phi_x, \phi_y \right), \tag{5.14}$$

as was pointed out by Bardi and Osher [12]. This is the canonical monotone scheme. Defining our intervals I^x and I^y in the LLLF manner using only the values of $\phi_x^\pm$ and $\phi_y^\pm$ at the grid node under consideration, we define ext_x and ext_y as follows. If $\phi_x^- < \phi_x^+$, then $\text{ext}_x H$ takes on the minimum value of H for all $\phi_x \in I^x$. If $\phi_x^- > \phi_x^+$, then $\text{ext}_x H$ takes on the maximum value of H for all $\phi_x \in I^x$. Otherwise, if $\phi_x^- = \phi_x^+$, then $\text{ext}_x H$ simply plugs $\phi_x^- (= \phi_x^+)$ into H for ϕ_x. Similarly, if $\phi_y^- < \phi_y^+$, then $\text{ext}_y H$ takes on the minimum value of H for all $\phi_y \in I^y$. If $\phi_y^- > \phi_y^+$, then $\text{ext}_y H$ takes on the maximum value of H for all $\phi_y \in I^y$. Otherwise, if $\phi_y^- = \phi_y^+$, then $\text{ext}_y H$ simply plugs $\phi_y^- (= \phi_y^+)$ into H for ϕ_y. In general, $\text{ext}_x \text{ext}_y H \neq \text{ext}_y \text{ext}_x H$, so different versions of Godunov's method are obtained depending on the order of operations. However, in many cases, including when H is separable, $\text{ext}_x \text{ext}_y H = \text{ext}_y \text{ext}_x H$ so this is not an issue.

Although Godunov's method can sometimes be difficult to implement, there are times when it is straightforward. Consider equation (3.2) for motion in an externally generated velocity field. Here, we can consider the x and y directions independently, since H is separable with $\text{ext}_x \text{ext}_y H = \text{ext}_x(u\phi_x) + \text{ext}_y(v\phi_y)$. If $\phi_x^- < \phi_x^+$, we want the minimum value of $u\phi_x$. Thus, if $u > 0$, we use ϕ_x^-, and if $u < 0$, we use ϕ_x^+. If $u = 0$, we obtain $u\phi_x = 0$ regardless of the choice of ϕ_x. On the other hand, if $\phi_x^- > \phi_x^+$, we want the maximum value of $u\phi_x$. Thus, if $u > 0$, we use ϕ_x^-, and if $u < 0$, we use ϕ_x^+. Again, $u = 0$ gives $u\phi_x = 0$. Finally, if $\phi_x^- = \phi_x^+$, then $u\phi_x$ is uniquely determined. This can be summarized as follows. If $u > 0$, use ϕ_x^-; if $u < 0$, use ϕ_x^+; and if $u = 0$, set $u\phi_x = 0$. This is identical to the standard upwind differencing method described in Chapter 3. That is, for motion in an externally generated velocity field, Godunov's method is identical to simple upwind differencing.

6

Motion in the Normal Direction

6.1 The Basic Equation

In this chapter we discuss the motion of an interface under an internally generated velocity field for constant motion in the normal direction. This velocity field is defined by $\vec{V} = a\vec{N}$ or $V_n = a$, where a is a constant. The corresponding level set equation (i.e., equation (4.4)) is

$$\phi_t + a|\nabla\phi| = 0, \qquad (6.1)$$

where a can be of either sign. When $a > 0$ the interface moves in the normal direction, and when $a < 0$ the interface moves opposite the normal direction. When $a = 0$ this equation reduces to the trivial $\phi_t = 0$, where ϕ is constant for all time. Figure 6.1 shows the evolution of a star-shaped interface as it moves normal to itself in the outward direction.

When ϕ is a signed distance function, equation (6.1) reduces to $\phi_t = -a$, and the values of ϕ either increase or decrease, depending on the sign of a. Forward Euler time discretization of this equation gives $\phi^{n+1} = \phi^n - a\triangle t$. When $a > 0$, the $\phi = 0$ isocontour becomes the $\phi = -a\triangle t$ isocontour after one time step. Similarly, the $\phi = a\triangle t$ isocontour becomes the $\phi = 0$ isocontour. That is, the $\phi = 0$ isocontour moves $a\triangle t$ units forward in the normal direction to the old position of the old $\phi = a\triangle t$ isocontour. The interface is moving in the normal direction with speed a. Taking the gradient of this forward Euler time stepping gives $\nabla\phi^{n+1} = \nabla\phi^n - \nabla(a\triangle t)$. Since $a\triangle t$ is spatially constant, $\nabla(a\triangle t) = 0$, implying that $\nabla\phi^{n+1} = \nabla\phi^n$.

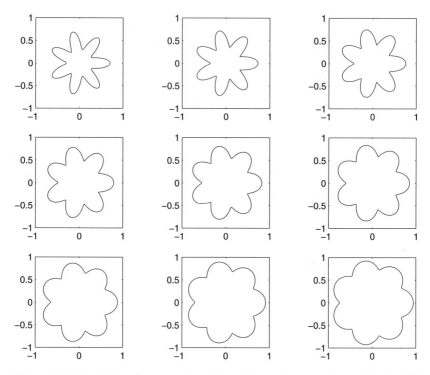

Figure 6.1. Evolution of a star-shaped interface as it moves normal to itself in the outward direction.

Thus, if ϕ^n is initially a signed distance function (with $|\nabla\phi^n| = 1$), it stays a distance function (with $|\nabla\phi| = 1$) for all time.

When the initial data constitute a signed distance function, forward Euler time stepping reduces to solving the ordinary differential equation $\phi_t = -a$ independently at every grid point. Since a is a constant, this forward Euler time stepping gives the exact solution up to round-off error (i.e., there is no truncation error). For example, consider a point where $\phi = \phi_o > 0$, which is ϕ_o units from the interface. In $\triangle t$ units of time the interface will approach $a\triangle t$ spatial units closer, changing the value of this point to $\phi_o - a\triangle t$, which is exactly the forward Euler time update of this point. The exact interface crossing time can be identified for all points by solving $\phi_o - at = 0$ to get $t = \phi_o/a$. (Similar arguments hold when $a < 0$, except that the interface moves in the opposite direction.)

Here, we see the power of signed distance functions. When ϕ_o is a signed distance function, we can write down the exact solution of equation (6.1) as $\phi(t) = \phi_o - at$. On the other hand, when ϕ_o is not a signed distance function, equation (6.1) needs to be solved numerically by treating it as a Hamilton-Jacobi equation, as discussed in the last chapter.

6.2 Numerical Discretization

For instructive purposes, suppose we plug $\vec{V} = a\vec{N}$ into equation (4.2) and try a simple upwind differencing approach. That is, we will attempt to discretize

$$\phi_t + \left(\frac{a\phi_x}{|\nabla\phi|}, \frac{a\phi_y}{|\nabla\phi|}, \frac{a\phi_z}{|\nabla\phi|} \right) \cdot \nabla\phi = 0 \qquad (6.2)$$

with simple upwinding. Consider the first spatial term $a\phi_x|\nabla\phi|^{-1}\phi_x$, where $a\phi_x|\nabla\phi|^{-1}$ is the "velocity" in the x-direction. Since upwinding is based only on the sign of the velocity, we can ignore the positive $|\nabla\phi|$ denominator, assuming temporarily that it is nonzero. Then the sign of $a\phi_x$ can be used to decide whether ϕ_x^- or ϕ_x^+ should be used to approximate ϕ_x. When ϕ_x^- and ϕ_x^+ have the same sign, it does not matter which of these is plugged into $a\phi_x$, since only the sign of this term determines whether we use ϕ_x^- or ϕ_x^+. For example, suppose $a > 0$. Then when $\phi_x^- > 0$ and $\phi_x^+ > 0$, $a\phi_x > 0$ and ϕ_x^- should be used in equation (6.2) everywhere ϕ_x appears, including the velocity term. On the other hand, when $\phi_x^- < 0$ and $\phi_x^+ < 0$, $a\phi_x < 0$ and ϕ_x^+ should be used to approximate ϕ_x.

This simple upwinding approach works well as long as ϕ_x^- and ϕ_x^+ have the same sign, but consider what happens when they have different signs. For example, when $\phi_x^- < 0$ and $\phi_x^+ > 0$, $a\phi_x^- < 0$ (still assuming $a > 0$), indicating that ϕ_x^+ should be used, while $a\phi_x^+ > 0$, indicating that ϕ_x^- should be used. This situation corresponds to a "V"-shaped region where each side of the "V" should move outward. The difficulty in approximating ϕ_x arises because we are in the vicinity of a sonic point, where $\phi_x = 0$. The LLLF interval defined by ϕ_x^- and ϕ_x^+ includes this sonic point since ϕ_x^- and ϕ_x^+ differ in sign. We have to take care to ensure that the expansion takes place properly. A similar problem occurs when $\phi_x^- > 0$ and $\phi_x^+ < 0$. Here $a\phi_x^- > 0$, indicating that ϕ_x^- should be used, while $a\phi_x^+ < 0$, indicating that ϕ_x^+ should be used. This upside-down "V" is shaped like a carrot (or hat) and represents the coalescing of information similar to a shock wave. Once again caution is needed to ensure that the correct solution is obtained.

Simple upwinding breaks down when ϕ_x^- and ϕ_x^+ differ in sign. Let us examine how the Roe-Fix method works in this case. In order to do this, we need to consider the Hamilton-Jacobi form of the equation, i.e., equation (6.1). Here $H_1 = a\phi_x|\nabla\phi|^{-1}$, implying that the simple velocity $\vec{V} = a\vec{N}$ we used in equation (6.2) was the correct expression to look at for upwinding. The sign of H_1 is independent of the y and z directions, depending only on $a\phi_x$. If both ϕ_x^- and ϕ_x^+ have the same sign, we choose one or the other depending on the sign of H_1 as in the usual upwinding. However, unlike simple upwinding, Roe-Fix gives a consistent method for treating the case where ϕ_x^- and ϕ_x^+ differ in sign. In that instance we are in the vicinity of a sonic point and switch to the LLF method, adding some numerical dissipation to the scheme in order to obtain the correct vanishing viscosity solution. The RF scheme treats the ambiguity associated with upwinding near sonic points by using central differencing plus some artificial viscosity.

Recall that numerical dissipation can smear out fine solution details on coarse grids. In order to avoid as much numerical smearing as possible, we have proposed five different versions (LF, SLF, LLF, SLLF, and LLLF) of the central differencing plus artificial viscosity approach to solving Hamilton-Jacobi problems. While the RF method is a better alternative, it too resorts to artificial dissipation in the vicinity of sonic points where ambiguities occur. In order to avoid the addition of artificial dissipation, one must resort to the Godunov scheme.

Let us examine the Godunov method in detail. Again, assume $a > 0$. If ϕ_x^- and ϕ_x^+ are both positive, then ext_x minimizes H when $\phi_x^- < \phi_x^+$ and maximizes H when $\phi_x^- > \phi_x^+$. In either case, we choose ϕ_x^- consistent with upwinding. Similarly, when ϕ_x^- and ϕ_x^+ are both negative, ext_x minimizes H when $\phi_x^- < \phi_x^+$ and maximizes H when $\phi_x^- > \phi_x^+$. Again, ϕ_x^+ is chosen in both instances consistent with upwinding. Now consider the "V"-shaped case where $\phi_x^- < 0$ and $\phi_x^+ > 0$, indicating an expansion. Here $\phi_x^- < \phi_x^+$, so Godunov's method minimizes H, achieving this minimum by setting $\phi_x = 0$. This implies that a region of expansion should have a locally flat ϕ with $\phi_x = 0$. Instead of adding numerical dissipation to hide the problem, Godunov's method chooses the most meaningful solution. Next, consider the case where $\phi_x^- > 0$ and $\phi_x^+ < 0$, indicating coalescing characteristics. Here $\phi_x^- > \phi_x^+$, so Godunov's method maximizes H, achieving this maximum by setting ϕ_x equal to the larger in magnitude of ϕ_x^- and ϕ_x^+. In this shock case, information is coming from both directions, and the grid point feels the effects of the information that gets there first. The velocities are characterized by $H_1 = a\phi_x|\nabla\phi|^{-1}$, and the side with the fastest speed arrives first. This is determined by taking the larger in magnitude of ϕ_x^- and ϕ_x^+. Again, Godunov's method chooses the most meaningful solution, avoiding artificial dissipation.

Godunov's method for equation (6.1) can be summarized as follows for both positive and negative a. If $a\phi_x^-$ and $a\phi_x^+$ are both positive, use $\phi_x = \phi_x^-$. If $a\phi_x^-$ and $a\phi_x^+$ are both negative, use $\phi_x = \phi_x^+$. If $a\phi_x^- \leq 0$ and $a\phi_x^+ \geq 0$, treat the expansion by setting $\phi_x = 0$. If $a\phi_x^- \geq 0$ and $a\phi_x^+ \leq 0$, treat the shock by setting ϕ_x to either ϕ_x^- or ϕ_x^+, depending on which gives the largest magnitude for $a\phi_x$. Note that when $\phi_x^- = \phi_x^+ = 0$ both of the last two cases are activated, and both consistently give $\phi_x = 0$. We also have the following elegant formula due to Rouy and Tourin [139]:

$$\phi_x^2 \approx \max\left(\max(\phi_x^-,0)^2, \min(\phi_x^+,0)^2\right) \tag{6.3}$$

when $a > 0$, and

$$\phi_x^2 \approx \max\left(\min(\phi_x^-,0)^2, \max(\phi_x^+,0)^2\right) \tag{6.4}$$

when $a < 0$.

6.3 Adding a Curvature-Dependent Term

Most flames burn with a speed in the normal direction plus extra heating and cooling effects due to the curvature of the front. This velocity field can be modeled by setting $V_n = a - b\kappa$ in the level set equation (4.4) to obtain

$$\phi_t + a|\nabla\phi| = b\kappa|\nabla\phi|; \tag{6.5}$$

which has both hyperbolic and parabolic terms. The hyperbolic $a|\nabla\phi|$ term can be discretized as outlined above using Hamilton-Jacobi techniques, while the parabolic $b\kappa|\nabla\phi|$ term can be independently discretized using central differencing as described in Chapter 4.

Once both terms have been discretized, either forward Euler or RK time discretization can be used to advance the front forward in time. The combined CFL condition for equations that contain both hyperbolic Hamilton-Jacobi terms and parabolic terms is given by

$$\triangle t \left(\frac{|H_1|}{\triangle x} + \frac{|H_2|}{\triangle y} + \frac{|H_3|}{\triangle z} + \frac{2b}{(\triangle x)^2} + \frac{2b}{(\triangle y)^2} + \frac{2b}{(\triangle z)^2} \right) < 1, \tag{6.6}$$

as one might have guessed from equation (4.12).

6.4 Adding an External Velocity Field

Equation (6.5) models a flame front burning through a material at rest, but does not account for the velocity of the unburnt material. A more general equation is

$$\phi_t + \vec{V} \cdot \nabla\phi + a|\nabla\phi| = b\kappa|\nabla\phi|, \tag{6.7}$$

since it includes the velocity $\vec{V}$ of the unburnt material. This equation combines an external velocity field with motion normal to the interface and motion by mean curvature. It is the most general form of the G-equation for burning flame fronts; see Markstein [110]. As in equation (6.5), the parabolic term on the right-hand side can be independently discretized with central differencing. The hyperbolic Hamilton-Jacobi part of this equation consists of two terms, $\vec{V} \cdot \nabla\phi$ and $a|\nabla\phi|$. Figure 6.2 shows the evolution of a star-shaped interface under the influence of both an externally given rigid-body rotation (a $\vec{V} \cdot \nabla\phi$ term) and a self-generated motion outward normal to the interface (an $a|\nabla\phi|$ term).

In order to discretize the Hamilton-Jacobi part of equation (6.7), we first identify the partial derivatives of H, i.e., $H_1 = u + a\phi_x|\nabla\phi|^{-1}$ and $H_2 = v + a\phi_y|\nabla\phi|^{-1}$. The first term in H_1 represents motion of the interface as it is passively advected in the external velocity field, while the second term represents the self-generated velocity of the interface as it moves normal to itself. If u and $a\phi_x$ have the same sign for both ϕ_x^- and ϕ_x^+, then the

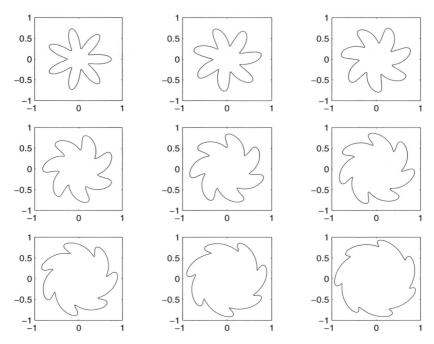

Figure 6.2. A star-shaped interface being advected by a rigid body rotation as it moves outward normal to itself.

background velocity and the self-generated velocity in the normal direction are both moving the front in the same direction, and the upwind direction is easy to determine. For example, if a, ϕ_x^-, and ϕ_x^+ are all positive, the second term in H_1 indicates that the self-generated normal velocity is moving the front to the right. Additionally, when $u > 0$ the external velocity is also moving the front to the right. In this case, both the RF scheme and the Godunov scheme set $\phi_x = \phi_x^-$.

It is more difficult to determine what is happening when u and $a\phi_x|\nabla\phi|^{-1}$ disagree in sign. In this case the background velocity is moving the front in one direction while the self-generated normal velocity is moving the front in the opposite direction. In order to upwind, we must determine which of these two terms dominates. It helps if ϕ is a signed distance function, since we obtain the simplified $H_1 = u + a\phi_x$. If H_1 is positive for both ϕ_x^- and ϕ_x^+, then both RF and Godunov set $\phi_x = \phi_x^-$. If H_1 is negative for both ϕ_x^- and ϕ_x^+, then both RF and Godunov set $\phi_x = \phi_x^+$. If H_1 is negative for ϕ_x^- and positive for ϕ_x^+, we have an expansion. If $\phi_x^- < \phi_x^+$, Godunov's method chooses the minimum value for H. This relative extremum occurs when $H_1 = 0$ implying that we set $\phi_x = -u/a$. If $\phi_x^- > \phi_x^+$ Godunov's method chooses the maximum value for H, which is again obtained by setting $\phi_x = -u/a$. If H_1 is positive for ϕ_x^- and negative for ϕ_x^+, we have a shock. Godunov's method dictates setting ϕ_x equal to the value of ϕ_x^- or ϕ_x^+ that gives the value of H_1 with the largest magnitude.

When ϕ is not a signed distance function, the above simplifications cannot be made. In general, $H_1 = u + a\phi_x |\nabla \phi|^{-1}$, and we need to consider not only I^x, but also the values of ϕ_y and ϕ_z in I^y and I^z, respectively. This can become rather complicated quite quickly. In fact, even the RF method can become quite complicated in this case, since it is hard to tell when sonic points are nearby and when they are not. In situations like this, the LLF scheme is ideal, since one merely uses both the values of ϕ_x^- and ϕ_x^+ along with some artificial dissipation setting α as dictated by equation (5.12). At first glance, equation (5.12) might seem complicated to evaluate; e.g., one has to determine the maximum value of $|H_1|$. However, since α is just a dissipation coefficient, we can safely overestimate α and pay the price of slightly more artificial dissipation. In contrast, it is hard to predict how certain approximations will affect the Godunov scheme. One way to approximate α is to separate H_1 into parts, i.e., using $|H_1| < |u| + |a\phi_x||\nabla \phi|^{-1}$ to treat the first and second terms independently. Also, when ϕ is approximately a signed distance, we can look at $|H_1| = |u + a\phi_x|$. This function is easy to maximize, since the maximum occurs at either ϕ_x^- or ϕ_x^+ and the y and z spatial directions play no role.

7
Constructing Signed Distance Functions

7.1 Introduction

As we have seen, a number of simplifications can be made when ϕ is a signed distance function. For this reason, we dedicate this chapter to numerical techniques for constructing approximate signed distance functions. These techniques can be applied to the initial data in order to initialize ϕ to a signed distance function.

As the interface evolves, ϕ will generally drift away from its initialized value as signed distance. Thus, the techniques presented in this chapter need to be applied periodically in order to keep ϕ approximately equal to signed distance. For a particular application, one has to decide how sensitive the relevant techniques are to ϕ's approximation of a signed distance function. If they are very sensitive, ϕ needs to be reinitialized to signed distance both accurately and often. If they are not sensitive, one can reinitialize with a lower-order accurate method on an occasional basis. However, even if a particular numerical approach doesn't seem to depend on how accurately ϕ approximates a signed distance function, one needs to remember that ϕ can develop noisy features and steep gradients that are not amenable to finite-difference approximations. For this reason, it is always advisable to reinitialize occasionally so that ϕ stays smooth enough to approximate its spatial derivatives with some degree of accuracy.

7.2   Reinitialization

In their seminal level set paper, Osher and Sethian [126] initialized their numerical calculations with $\phi = 1 \pm d^2$, where d is the distance function and the "$\pm$" sign is negative in Ω^- and positive in Ω^+. Later, it became clear that the signed distance function $\phi = \pm d$, was a better choice for initializing ϕ. Mulder, Osher, and Sethian [115] demonstrated that initializing ϕ to a signed distance function results in more accurate numerical solutions than initializing ϕ to a Heaviside function. While it is obvious that better results can be obtained with smooth functions than nonsmooth functions, there are those who insist on using (slightly smeared out) Heaviside functions, or *color functions*, to track interfaces.

In [48], Chopp considered an application where certain regions of the flow had level sets piling up on each other, increasing the local gradient, while other regions of the flow had level sets separating from each other, flattening out ϕ. In order to reduce the numerical errors caused by both steepening and flattening effects, [48] introduced the notion that one should *reinitialize* the level set function periodically throughout the calculation. Since only the $\phi = 0$ isocontour has any meaning, one can stop the calculation at any point in time and reset the other isocontours so that ϕ is again initialized to a signed distance function. The most straightforward way of implementing this is to use a contour plotting algorithm to locate and discretize the $\phi = 0$ isocontour and then explicitly measure distances from it. Unfortunately, this straightforward reinitialization routine can be slow, especially if it needs to be done at every time step. In order to obtain reasonable run times, [48] restricted the calculations of the interface motion and the reinitialization to a small band of points near the $\phi = 0$ isocontour, producing the first version of the *local level set* method. We refer those interested in local level set methods to the more recent works of Adalsteinsson and Sethian [2] and Peng, Merriman, Osher, Zhao, and Kang [130].

The concept of frequent reinitialization is a powerful numerical tool. In a standard numerical method, one starts with initial data and proceeds forward in time, assuming that the numerical solution stays well behaved until the final solution is computed. With reinitialization, we have a less-stringent assumption, since only our $\phi = 0$ isocontour needs to stay well behaved. Any problems that creep up elsewhere are wiped out when the level set is reinitialized. For example, Merriman, Bence, and Osher [114] proposed numerical techniques that destroy the nice properties of the level set function and show that poor numerical solutions are obtained using these degraded level set functions. Then they show that periodic reinitialization to a signed distance function repairs the damage, producing high-quality numerical results. Numerical techniques need to be effective only for the $\phi = 0$ isocontour, since the rest of the implicit surface can be repaired by reinitializing ϕ to a signed distance function. This greatly increases flexibility in algorithm design, since difficulties away from the $\phi = 0$ isocontour can be ignored.

7.3 Crossing Times

One of the difficulties associated with the direct computation of signed distance functions is locating and discretizing the interface. This can be avoided in the following fashion. Consider a point $\vec{x} \in \Omega^+$. If $\vec{x}$ does not lie on the interface, we wish to know how far from the interface it is so that we can set $\phi(\vec{x}) = +d$. If we move the interface in the normal direction using equation (6.1) with $a = 1$, the interface eventually crosses over $\vec{x}$, changing the local value of ϕ from positive to negative. If we keep a time history of the local values of ϕ at $\vec{x}$, we can find the exact time when ϕ was equal to zero using interpolation in time. This is the time it takes the zero level set to reach the point $\vec{x}$, and we call that time t_o the *crossing time*. Since equation (6.1) moves the level set normal to itself with speed $a = 1$, the time it takes for the zero level set to reach a point $\vec{x}$ is equal to the distance the interface is from $\vec{x}$. That is, the crossing time t_o is equal to the distance d. For points $\vec{x} \in \Omega^-$, the crossing time is similarly determined using $a = -1$ in equation (6.1).

In a series of papers, [20], [97], and [100], Kimmel and Bruckstein introduced the notion of using crossing times in image-processing applications. For example, [100] used equation (6.1) with $a = 1$ to create *shape offsets*, which are distance functions with distance measured from the boundary of an image. The idea of using crossing times to solve some general Hamilton-Jacobi equations with Dirichlet boundary conditions was later generalized and rigorized by Osher [123].

7.4 The Reinitialization Equation

In [139], Rouy and Tourin proposed a numerical method for solving $|\nabla \phi| = f(\vec{x})$ for a spatially varying function f derived from the intensity of an image. In the trivial case of $f(\vec{x}) = 1$, the solution ϕ is a signed distance function. They added $f(\vec{x})$ to the right-hand side of equation (6.1) as a source term to obtain

$$\phi_t + |\nabla \phi| = f(\vec{x}), \qquad (7.1)$$

which is evolved in time until a steady state is reached. At *steady state*, the values of ϕ cease to change, implying that $\phi_t = 0$. Then equation (7.1) reduces to $|\nabla \phi| = f(\vec{x})$, as desired. Since only the steady-state solution is desired, [139] used an accelerated iteration method instead of directly evolving equation (7.1) forward in time.

Equation (7.1) propagates information in the normal direction, so information flows from smaller values of ϕ to larger values of ϕ. This equation is of little use in reinitializing the level set function, since the interface location will be influenced by the negative values of ϕ. That is, the $\phi = 0$ isocontour is not guaranteed to stay fixed, but will instead move around as it is influenced by the information flowing in from the negative values of ϕ. One way to avoid this is to compute the signed distance function for all the grid points adjacent to the interface by hand. Then

$$\phi_t + |\nabla\phi| = 1 \qquad (7.2)$$

can be solved in Ω^+ to update ϕ based on those grid points adjacent to the interface. That is, the grid points adjacent to the interface are not updated, but instead used as boundary conditions. Since there is only a single band of initialized grid cells on each side of the interface, one cannot apply higher-order accurate methods such as HJ WENO. However, if a two-grid-cell-thick band is initialized in Ω^- (in addition to the one-grid-cell-thick band in Ω^+), the total size if the band consists of three grid cells and the HJ WENO scheme can then be used. Alternatively, one could initialize a three-grid-cell-thick band of boundary conditions in Ω^- and use these to update every point in Ω^+ including those adjacent to the interface. Similarly, a three grid cell thick band of boundary conditions can be initialized in Ω^+ and used to update the values of ϕ in Ω^- by solving

$$\phi_t - |\nabla\phi| = -1 \qquad (7.3)$$

to steady state. Equations (7.2) and (7.3) reach steady state rather quickly, since they propagate information at speed 1 in the direction normal to the interface. For example, if $\Delta t = 0.5\Delta x$, it takes only about 10 time steps to move information from the interface to 5 grid cells away from the interface.

In [160], Sussman, Smereka, and Osher put all this together into a *reinitialization equation*

$$\phi_t + S(\phi_o)(|\nabla\phi| - 1) = 0, \qquad (7.4)$$

where $S(\phi_o)$ is a sign function taken as 1 in Ω^+, -1 in Ω^-, and 0 on the interface, where we want ϕ to stay identically equal to zero. Using this equation, there is no need to initialize any points near the interface for use as boundary conditions. The points near the interface in Ω^+ use the points in Ω^- as boundary conditions, while the points in Ω^- conversely look at those in Ω^+. This circular loop of dependencies eventually balances out, and a steady-state signed distance function is obtained. As long as ϕ is relatively smooth and the initial data are somewhat balanced across the interface, this method works rather well. Unfortunately, if ϕ is not smooth or ϕ is much steeper on one side of the interface than the other, circular dependencies on initial data can cause the interface to move incorrectly from its initial starting position. For this reason, [160] defined $S(\phi_o)$ using the initial values of ϕ (denoted by ϕ_o) so that the domain of dependence does not change if the interface incorrectly crosses over a grid point. This was addressed directly by Fedkiw, Aslam, Merriman, and Osher in the appendix of [63], where incorrect interface crossings were identified as sign changes in the nodal values of ϕ. These incorrect interface crossings were rectified by putting the interface back on the correct side of a grid point $\vec{x}$, setting $\phi(\vec{x}) = \pm\epsilon$, where $\pm\epsilon$ is a small number with the appropriate sign.

In discretizing equation (7.4), the $S(\phi_o)|\nabla\phi|$ term is treated as motion in the normal direction as described in Chapter 6. Here $S(\phi_o)$ is constant for all time and can be thought of as a spatially varying "a" term. Numerical tests indicate that better results are obtained when $S(\phi_o)$ is numerically smeared out, so [160] used

$$S(\phi_o) = \frac{\phi_o}{\sqrt{\phi_o^2 + (\triangle x)^2}} \tag{7.5}$$

as a numerical approximation. Later, Peng, Merriman, Osher, Zhao, and Kang [130] suggested that

$$S(\phi) = \frac{\phi}{\sqrt{\phi^2 + |\nabla\phi|^2(\triangle x)^2}} \tag{7.6}$$

was a better choice, especially when the initial ϕ_o was a poor estimate of signed distance, i.e., when $|\nabla\phi_o|$ was far from 1. In equation (7.6), it is important to update $S(\phi)$ continually as the calculation progresses so that the $|\nabla\phi|$ term has the intended effect. In contrast, equation (7.5) is evaluated only once using the initial data. Numerical smearing of the sign function decreases its magnitude, slowing the propagation speed of information near the interface. This probably aids the balancing out of the circular dependence on the initial data as well, since it produces characteristics that do not look as far across the interface for their information. We recommend using Godunov's method for discretizing the hyperbolic $S(\phi_o)|\nabla\phi|$ term. After finding a numerical approximation to $S(\phi_o)|\nabla\phi|$, we combine it with the remaining $S(\phi_o)$ source term at each grid point and update the resulting quantity in time with a Runge-Kutta method.

Ideally, the interface remains stationary during the reinitialization procedure, but numerical errors will tend to move it to some degree. In [158], Sussman and Fatemi suggested an improvement to the standard reinitialization procedure. Since their application of interest was two-phase incompressible flow, they focused on preserving the amount of material in each cell, i.e., preserving the area (volume) in two (three) spatial dimensions. If the interface does not move during reinitialization, the area is preserved. On the other hand, one can preserve the area while allowing the interface to move, implying that their proposed constraint is weaker than it should be. In [158] this constraint was applied locally, requiring that the area be preserved individually in each cell. Instead of using the exact area, the authors used equation (1.15) with $f(\vec{x}) = 1$ to approximate the area in each cell as

$$A_{i,j} = \int_{\Omega_{i,j}} H(\phi)\, d\vec{x}, \tag{7.7}$$

where $\Omega_{i,j}$ is an individual grid cell and H is the smeared-out Heaviside function in equation (1.22). In both [158] and the related [159] by Sussman, Fatemi, Smereka, and Osher this local constraint was shown to significantly improve the results obtained with the HJ ENO scheme. However, this local constraint method has not yet been shown to improve upon the results obtained with the significantly more accurate HJ WENO scheme. The concern is that the HJ WENO scheme might be so accurate that the approximations made by [158] could lower the accuracy of the method.

This local constraint is implemented in [158] by the addition of a correction term to the right-hand side of equation (7.4),

$$\phi_t + S(\phi_o)(|\nabla\phi| - 1) = \lambda\delta(\phi)|\nabla\phi|, \tag{7.8}$$

where the multidimensional delta function $\hat{\delta} = \delta(\phi)|\nabla\phi|$ from equation (1.19) is used, since the modifications are needed only near the interface where $A_{i,j}$ is not trivially equal to either zero or the volume of $\Omega_{i,j}$. The constraint that $A_{i,j}$ in each cell not change, i.e., $(A_{i,j})_t = 0$, is equivalent to

$$\int_{\Omega_{i,j}} H'(\phi)\phi_t\, d\vec{x} = 0, \tag{7.9}$$

or

$$\int_{\Omega_{i,j}} \delta(\phi)\left(-S(\phi_o)(|\nabla\phi| - 1) + \lambda\delta(\phi)|\nabla\phi|\right) d\vec{x} = 0, \tag{7.10}$$

using equation (7.8) and the fact that $H'(\phi) = \delta(\phi)$ (see equation (1.18)). A separate $\lambda_{i,j}$ is defined in each cell using equation (7.10) to obtain

$$\lambda_{i,j} = -\frac{\int_{\Omega_{i,j}} \delta(\phi)\left(-S(\phi_o)(|\nabla\phi| - 1)\right) d\vec{x}}{\int_{\Omega_{i,j}} \delta^2(\phi)|\nabla\phi|\, d\vec{x}}, \tag{7.11}$$

or

$$\lambda_{i,j} = -\frac{\int_{\Omega_{i,j}} \delta(\phi)\left(\frac{\phi^{n+1}-\phi^n}{\Delta t}\right) d\vec{x}}{\int_{\Omega_{i,j}} \delta^2(\phi)|\nabla\phi|\, d\vec{x}}, \tag{7.12}$$

where equation (7.4) is used to compute ϕ^{n+1} from ϕ^n. In summary, first equation (7.4) is used to update ϕ^n in time using, for example, an RK method. Then equation (7.12) is used to compute a $\lambda_{i,j}$ for each grid cell. Sussman and Fatemi in [158] used a nine-point quadrature formula to evaluate the integrals in two spatial dimensions. Finally, the initial guess for ϕ^{n+1} obtained from equation (7.4) is replaced with a corrected $\phi^{n+1} + \Delta t\lambda\delta(\phi)|\nabla\phi|$. It is shown in [158] that this specific discretization exactly cancels out a first order error term in the previous formulation. This procedure is very similar to that used by Rudin, Osher, and Fatemi [142] as a continuous in time gradient projection method. In [142] a set of constraints needs to be preserved under evolution, while in [158] the evolution is not inherited from gradient descent on a functional to be optimized.

Reinitialization is still an active area of research. Recently, Russo and Smereka [143] introduced yet another method for computing the signed distance function. This method was designed to keep the stencil from incorrectly looking across the interface at values that should not influence it, essentially removing the balancing act between the interdependent initial data across the interface. Their idea replaces equation (7.4) with a combination of equations (7.2) and (7.3) along with interpolation to find the interface location. In [143] marked improvement was shown in using low-order HJ ENO schemes, but the authors did not address whether they can obtain improved results over the recommended HJ WENO discretization of equation (7.4). Moreover, implementing a high-order accurate version of the scheme in [143] requires a number of ghost cells, as discussed above.

7.5 The Fast Marching Method

In the crossing-time approach to constructing signed distance functions, the zero isocontour moves in the normal direction, crossing over grid points at times equal to their distance from the interface. In this fashion, each grid point is updated as the zero isocontour crosses over it. Here we discuss a discrete algorithm that mimics this approach by marching out from the interface calculating the signed distance function at each grid point.

Suppose that all the grid points adjacent to the interface are initialized with the exact values of signed distance. We will discuss methods for initializing this band of cells later. Starting from this initial band, we wish to march outward, updating each grid point with the appropriate value of signed distance. Here we describe the algorithm for marching in the normal direction to construct the positive distance function, noting that the method for marching in the direction opposite the normal to construct the negative distance function is applied in the same manner. In fact, if the values in the initial band are multiplied by -1, the positive distance function construction can be used to find positive distance values in Ω^- that can then be multiplied by -1 to obtain appropriate negative distance values in this region.

In order to march out from the initial band, constructing the distance function as we go, we need to decide which grid point to update first. This should be the one that the zero isocontour would cross first in the crossing-time method, i.e., the grid point with the smallest crossing time or smallest value of distance. So, for each grid point adjacent to the band, we calculate a tentative value for the distance function. This is done using only the values of ϕ that have already been accepted into the band; i.e., tentative values do not use other tentative values in this calculation. Then we choose the point with the smallest tentative value to add to the band of accepted grid points. Since the signed distance function is created with characteristics that flow out of the interface in the normal direction, this chosen point does not depend on any of the other tentative grid points that will have larger values of distance. Thus, the tentative value of distance assigned to this grid point is an acceptable approximation of the signed distance function.

Now that the band of accepted values has been increased by one, we repeat the process. Most of the grid points in the tentative band already have good tentative approximations to the distance function. Only those adjacent to the newly added point need modification. Adjacent tentative grid points need their tentative values updated using the new information gained by adding the chosen point to the band. Any other adjacent grid point that did not yet have a tentative value needs to have a tentative value assigned to it using the values in the band of accepted points. Then we choose the smallest tentative value, add it to the band, and repeat the algorithm. Eventually, every grid point in Ω^+ gets added to the band, completing the process. As noted above, the grid points in Ω^- are updated with a similar process.

The slowest part of this algorithm is the search through all the tentative grid points to find the one with the smallest value. This search can be accelerated using a binary tree to store all the tentative values. The tree is organized so that each point has a tentative distance value smaller than the two points located below it in the tree. This means that the smallest tentative point is always conveniently located at the top of the tree. New points are added to the bottom of the tree, where we note that the method works better if the tree is kept balanced. If the newly added point has a smaller value of distance than the point directly above it, we exchange the location of these two points in the tree. Recursively, this process is repeated, and the newly added point moves up the tree until it either sits below a point with a smaller tentative distance value or it reaches the top of the tree. We add points to the bottom of the tree as opposed to the top, since newly added points tend to be farther from the interface with larger distance values than those already in the tree. This means that fewer comparisons are generally needed for a newly added point to find an appropriate location in the tree.

The algorithm proceeds as follows. Remove the point from the top of the tree and add it to the band. The vacated space in the tree is filled with the smaller of the two points that lie below it. Recursively, the holes opened up by points moving upward are filled with the smaller of the two points that lie below until the bottom of the tree is reached. Next, any tentative values adjacent to the added point are updated by changing their tentative values. These then need to be moved up or down the tree in order to preserve the ordering based on tentative distance values. In general, tentative values should only decrease, implying that the updated point may have to be moved up the tree. However, numerical error could occasionally cause a tentative distance value to increase (if only by round-off error) in which case the point may need to be moved down lower in the tree. Tentative distance values are calculated at each new adjacent grid point that was not already in the tree, and these points are added to the tree. The algorithm is $O(N \log N)$, where N is the total number of grid points.

This algorithm was invented by Tsitsiklis in a pair of papers, [166] and [167]. The most novel part of the algorithm is the extension of Dijkstra's method for computing the taxicab metric to an algorithm for computing Euclidean distance. In these papers, $\phi_{i,j,k}$ is chosen to be as small as possible by obtaining the correct solution in the sense of first arrival time. First, each quadrant is independently considered to find the characteristic direction $\vec{\theta} = (\theta_1, \theta_2, \theta_3)$, where each $\theta_s > 0$ and $\sum \theta_s = 1$, that gives the smallest value for $\phi_{i,j,k}$. Then the values from all the quadrants are compared, and the smallest of these is chosen as the tentative guess for $\phi_{i,j,k}$. That is, the characteristic direction is found by first finding the best candidate in each quadrant and then comparing these (maximum of eight) candidates to find the best global candidate.

In [166] and [167], the minimum value of $\phi_{i,j,k}$ in a particular quadrant is found by minimizing

$$\phi_{i,j,k} = \tau(\vec{\theta}) + \theta_1\phi_1 + \theta_2\phi_2 + \theta_3\phi_3 \tag{7.13}$$

over all directions $\vec{\theta}$, where

$$\tau(\vec{\theta}) = \sqrt{(\theta_1 \triangle x_1)^2 + (\theta_2 \triangle x_2)^2 + (\theta_3 \triangle x_3)^2} \tag{7.14}$$

is the distance traveled and $\sum \theta_s \phi_s$ is the starting point in the particular quadrant. There are eight possible quadrants, with starting points determined by $\phi_1 = \phi_{i\pm1,j,k}$, $\phi_2 = \phi_{i,j\pm1,k}$, and $\phi_3 = \phi_{i,j,k\pm1}$. If any of the arms of the stencil is not in the band of updated points, this arm is simply ignored. In the minimization formalism, this is accomplished by setting points outside the updated band to ∞ and using the convention that $0 \cdot \infty = 0$. This sets the corresponding $\theta_s \phi_s$ term in equation (7.13) to zero for any $\phi = \infty$ not in the band of updated points simply by setting $\theta_s = 0$, i.e., by ignoring that direction of the stencil.

A Lagrange multiplier λ is added to equation (7.13) to obtain

$$\phi_{i,j,k} = \tau(\vec{\theta}) + \theta_1\phi_1 + \theta_2\phi_2 + \theta_3\phi_3 + \lambda\left(1 - \sum \theta_s\right), \tag{7.15}$$

where $1 - \sum \theta_s = 0$. For each θ_s, we take a partial derivative and set it to zero, obtaining

$$\frac{\theta_s(\triangle x_s)^2}{\tau(\vec{\theta})} + \phi_s - \lambda = 0 \tag{7.16}$$

in standard fashion. Solving equation (7.16) for each ϕ_s and plugging the results into equation (7.13) yields (after some cancellation) $\phi_{i,j,k} = \lambda$; i.e., λ is our minimum value. To find λ, we rewrite equation (7.16) as

$$\left(\frac{\lambda - \phi_s}{\triangle x_s}\right)^2 = \frac{\theta_s^2 (\triangle x_s)^2}{\tau(\vec{\theta})^2} \tag{7.17}$$

and sum over all spatial dimensions to obtain

$$\left(\frac{\lambda - \phi_1}{\triangle x_1}\right)^2 + \left(\frac{\lambda - \phi_2}{\triangle x_2}\right)^2 + \left(\frac{\lambda - \phi_3}{\triangle x_3}\right)^2 = 1, \tag{7.18}$$

using equation (7.14) to reduce the right-hand side of equation (7.18) to 1.

In summary, [166] and [167] compute the minimum value of $\phi_{i,j,k}$ in each quadrant by solving the quadratic equation

$$\left(\frac{\phi_{i,j,k} - \phi_1}{\triangle x}\right)^2 + \left(\frac{\phi_{i,j,k} - \phi_2}{\triangle y}\right)^2 + \left(\frac{\phi_{i,j,k} - \phi_3}{\triangle z}\right)^2 = 1. \tag{7.19}$$

Then the final value of $\phi_{i,j,k}$ is taken as the minimum over all the quadrants. Equation (7.19) is a first-order accurate approximation of $|\nabla \phi|^2 = 1$, i.e., the square of $|\nabla \phi| = 1$.

The final minimization over all the quadrants is straightforward. For example, with ϕ_2 and ϕ_3 fixed, the smaller value of $\phi_{i,j,k}$ is obtained as $\phi_1 = \min(\phi_{i-1,j,k}, \phi_{i+1,j,k})$, ruling out four quadrants. The same considerations apply to $\phi_2 = \min(\phi_{i,j-1,k}, \phi_{i,j+1,k})$ and $\phi_3 = \min(\phi_{i,j,k-1}, \phi_{i,j,k+1})$. Equation (7.19) is straightforward to solve using these definitions of ϕ_1, ϕ_2, and ϕ_3. This is equivalent to using either the forward difference or the backward difference to approximate each derivative of ϕ. If these definitions give $\phi_s = \infty$, than neither the forward nor the backward difference is defined since both the neighboring points in that spatial dimension are not in the accepted band. In this instance, we set $\theta_s = 0$, which according to equation (7.17) is equivalent to dropping the troublesome term out of equation (7.19) setting it to zero.

Each of ϕ_1, ϕ_2, and ϕ_3 can potentially be equal to ∞ if there are no neighboring accepted band points in the corresponding spatial dimension. If one of these quantities is equal to ∞, the corresponding term vanishes from equation (7.19) as we set the appropriate $\theta_s = 0$. Since there is always at least one adjacent point in the accepted band, at most two of the three terms can vanish, giving

$$\left(\frac{\phi_{i,j,k} - \phi_s}{\triangle x_s}\right)^2 = 1, \tag{7.20}$$

which can be solved to obtain $\phi_{i,j,k} = \phi_s \pm \triangle x_s$. The larger term, denoted by the "+" sign, is the one we use, since distance increases as the algorithm proceeds. When there are two nonzero terms, equation (7.19) becomes

$$\left(\frac{\phi_{i,j,k} - \phi_{s_1}}{\Delta x_{s_1}}\right)^2 + \left(\frac{\phi_{i,j,k} - \phi_{s_2}}{\Delta x_{s_2}}\right)^2 = 1, \tag{7.21}$$

where s_1 and s_2 represent different spatial dimensions. This quadratic equation can have zero, one, or two solutions. While this theoretically should not happen, it can be caused by poor initial data or numerical errors. Defining

$$P(\phi_{i,j,k}) = \left(\frac{\phi_{i,j,k} - \phi_{s_1}}{\Delta x_{s_1}}\right)^2 + \left(\frac{\phi_{i,j,k} - \phi_{s_2}}{\Delta x_{s_2}}\right)^2 \tag{7.22}$$

allows us to write $P(\max\{\phi_{s_1}, \phi_{s_2}\}) \leq 1$ as a necessary and sufficient condition to find an adequate solution $\phi_{i,j,k}$ of equation (7.21). If $P(\max\{\phi_{s_1}, \phi_{s_2}\}) > 1$, then $\phi_{i,j,k} < \max\{\phi_{s_1}, \phi_{s_2}\}$ if a solution $\phi_{i,j,k}$ exists. This implies that something is wrong (probably due to poor initial data or numerical error), since larger values of ϕ should not be contributing to smaller values. In order to obtain the best solution under the circumstances, we discard the term with the larger ϕ_s and proceed with equation (7.20). Otherwise, when $P(\max\{\phi_{s_1}, \phi_{s_2}\}) \leq 1$, equation (7.21) has two solutions, and we use the larger one, corresponding to the "+" sign in the quadratic formula. Similarly, when all three terms are present, we define

$$P(\phi_{i,j,k}) = \left(\frac{\phi_{i,j,k} - \phi_1}{\Delta x}\right)^2 + \left(\frac{\phi_{i,j,k} - \phi_2}{\Delta y}\right)^2 + \left(\frac{\phi_{i,j,k} - \phi_3}{\Delta z}\right)^2 \tag{7.23}$$

and take the larger solution, corresponding to the "+" sign, when $P(\max\{\phi_s\}) \leq 1$. Otherwise, when $P(\max\{\phi_s\}) > 1$ we omit the term with the largest ϕ_s and proceed with equation (7.22).

Consider the initialization of the grid points in the band about the interface. The easiest approach is to consider each of the coordinate directions independently. If ϕ changes sign in a coordinate direction, linear interpolation can be used to locate the interface crossing and determine a candidate value of ϕ. Then ϕ is initialized using the candidate with the smallest magnitude. Both this initialization routine and the marching algorithm itself are first-order accurate. For this reason, the reinitialization equation is often a better choice, since it is highly accurate in comparison. On the other hand, reinitialization is significantly more expensive and does not work well when ϕ is not initially close to signed distance. Thus, in many situations this optimal $O(N \log N)$ algorithm is preferable.

Although the method described above was originally proposed by Tsitsiklis in [166] and [167], it was later rediscovered by the level set community; see, for example, Sethian [148] and Helmsen, Puckett, Colella, and Dorr [85], where it is popularly referred to as the *fast marching method* (FMM). In [149], Sethian pointed out that higher-order accuracy could be achieved by replacing the first-order accurate forward and backward differences used by [166] and [167] in equation (7.19) with second-order accurate forward and backward differences whenever there are enough points in the updated band to evaluate these higher-order accurate differences. The second-order accurate versions of equations (1.3) and (1.4) are

$$\frac{\partial \phi}{\partial x} \approx \frac{\phi_{i+1} - \phi_i}{\Delta x} + \left(\frac{\Delta x}{2}\right) \left(\frac{\phi_{i+2} - 2\phi_{i+1} + \phi_i}{(\Delta x)^2}\right) = \frac{\phi_{i+2} - 4\phi_{i+1} + 3\phi_i}{2\Delta x}$$

$$(7.24)$$

and

$$\frac{\partial \phi}{\partial x} \approx \frac{\phi_i - \phi_{i-1}}{\Delta x} + \left(\frac{\Delta x}{2}\right) \left(\frac{\phi_i - 2\phi_{i-1} + \phi_{i-2}}{(\Delta x)^2}\right) = \frac{3\phi_i - 4\phi_{i-1} + \phi_{i-2}}{2\Delta x},$$

$$(7.25)$$

respectively. This lowers the local truncation error whenever more accepted band points are available. As pointed out in [149], higher-order accurate (higher than second order) forward and backward differences could be used as well. One difficulty with obtaining higher-order accurate solutions is that the initial band adjacent to the interface is constructed with first-order accurate linear interpolation. In [49], Chopp proposed using higher-order accurate interpolants to initialize the points adjacent to the interface. This leads to a set of equations that are not trivial to solve, and [49] used a variant of Newton iteration to find an approximate solution. When the iteration method (occasionally) fails, [49] uses the lower-order accurate linear interpolation to initialize the problematic points. Overall, the iteration scheme converges often enough to significantly improve upon the results obtained using the lower-order accurate linear interpolation everywhere.

8

Extrapolation in the Normal Direction

8.1 One-Way Extrapolation

In the last chapter we constructed signed distance functions by following characteristics that flow outward from the interface. Similar techniques can be used to propagate information in the direction of these characteristics. For example,

$$S_t + \vec{N} \cdot \nabla S = 0 \tag{8.1}$$

is a Hamilton-Jacobi equation (in S) that extrapolates S normal to the interface, i.e. so that S is constant on rays normal to the interface. Since $H(\nabla S) = \vec{N} \cdot \nabla S$, we can solve this equation with the techniques presented in Chapter 5 using $H_1 = n_1$, $H_2 = n_2$, and $H_3 = n_3$.

While central differencing can be used to compute the normal, it is usually advantageous to use upwind differencing here, since this equation is propagating information along these characteristics. At a given point $\vec{x}_{i,j,k}$ where the level set function is $\phi_{i,j,k}$, we determine ϕ_x by considering both $\phi_{i-1,j,k}$ and $\phi_{i+1,j,k}$. If either of these values is smaller than $\phi_{i,j,k}$, we use the minimum of these two values to compute a one-sided difference. On the other hand, if both of these values are larger than $\phi_{i,j,k}$, we set $\phi_x = 0$; noting that no S information should be flowing into this point along the x direction. After computing ϕ_y and ϕ_z in the same fashion, equation (1.2) can be used to define the normal direction.

Suppose that S is initially defined only in Ω^- and we wish to extend its values across the interface from Ω^- into Ω^+. Solving equation (8.1) in Ω^+

using the values of S in Ω^- as boundary conditions extrapolates S across the interface constant in the normal direction. This was done by Fedkiw, Aslam, Merriman, and Osher in [63] to solve multiphase flow problems with a ghost fluid method. We can extrapolate S in the opposite direction from Ω^+ to Ω^- by solving

$$S_t - \vec{N} \cdot \nabla S = 0 \qquad (8.2)$$

in Ω^- using the values of S in Ω^+ as boundary conditions. Of course, the upwind normal should be computed using the larger neighboring values of ϕ instead of the smaller neighboring values.

8.2   Two-Way Extrapolation

Just as we combined equations (7.2) and (7.3) to obtain equation (7.4), equations (8.1) and (8.2) can be combined to obtain

$$S_t + S(\phi)\vec{N} \cdot \nabla S = 0 \qquad (8.3)$$

to extrapolate S away from the interface. Here the upwind version of $\vec{N}$ is computed using smaller values of ϕ in Ω^+ and larger values of ϕ in Ω^-. Equation (8.3) can be applied to any value S that needs to be smoothed normal to the interface. For example, if this equation is solved independently for each component of the velocity field, i.e., $S = u$, $S = v$, and $S = w$, we obtain a velocity field that is constant normal to the interface. Velocity fields of this type have a tendency to preserve signed distance functions, as discussed by Zhao, Chan, Merriman, and Osher in [175]. This velocity extrapolation is also useful when the velocity is known only near the interface. For example, in [43], Chen, Merriman, Osher, and Smereka computed the velocity field for grid cells adjacent to the interface using local information from both sides of the interface. Then the velocity values in this band were held fixed, and equation (8.3) was used to extend each component of the velocity field outward from this initial band.

8.3   Fast Marching Method

As in the construction of signed distance functions, the fast marching method can be used to extrapolate S in an efficient manner. For example, consider a fast marching method alternative to equation (8.1). The normal is computed using the smaller neighboring values of ϕ (as above). The binary heap structure can be precomputed using all the points in Ω^+, as opposed to just using the points adjacent to the initialized band, since ϕ is already defined throughout Ω^+. Once the points are ordered, we choose the point with the smallest value of ϕ and compute an appropriate value for S. Then we find the next smallest value of ϕ, compute S, and continue in this fashion. The tentative values play no role, since we already know ϕ and thus the characteristic directions, i.e., assuming that ϕ is a signed distance function.

At each grid point, S is determined using the values of S at the neighboring points in a fashion dictated by the neighboring values of ϕ. Since S should be constant normal to the interface, we set $\vec{N} \cdot \nabla S = 0$. Or equivalently, since $\vec{N}$ and $\nabla \phi$ point in the same direction, we set $\nabla \phi \cdot \nabla S = 0$, where the derivatives in $\nabla \phi$ are computed in the fashion outlined above for computing normals. Then

$$\phi_x S_x + \phi_y S_y + \phi_z S_z = 0 \qquad (8.4)$$

is discretized using $\nabla \phi$ to determine the upwind direction. That is, we use $S_x = D^- S_{i,j,k}$ when $\phi_x > 0$ and $S_x = D^+ S_{i,j,k}$ when $\phi_x < 0$. When $\phi_x = 0$, the neighboring values of ϕ are larger than $\phi_{i,j,k}$, and no S information can be obtained from either of the neighboring nodes. In this case, we drop the S_x term from equation (8.4). The S_y and S_z terms are treated in a similar fashion. If ϕ is a signed distance function, this method works well. For more details on the fast marching alternative to equation (8.1), see Adalsteinsson and Sethian [1].

9
Particle Level Set Method

9.1 Eulerian Versus Lagrangian Representations

The great success of level set methods can in part be attributed to the role of curvature in regularizing the level set function such that the proper vanishing viscosity solution is obtained. It is much more difficult to obtain vanishing viscosity solutions with Lagrangian methods that faithfully follow the characteristics. For these methods one usually has to delete (or add) characteristic information "by hand" when a shock (or rarefaction) is detected. This ability of level set methods to identify and delete merging characteristics is clearly seen in a purely geometrically driven flow where a curve is advected normal to itself at constant speed, as shown in Figures 9.1 and 9.2. In the corners of the square, the flow field has merging characteristics that are appropriately deleted by the level set method. We demonstrate the difficulties associated with a Lagrangian calculation of this interface motion by initially seeding some marker particles interior to the interface, as shown in Figure 9.3 and passively advecting them with $\vec{x}_t = \vec{V}(\vec{x}, t)$; where the velocity field $\vec{V}(\vec{x}, t)$ is determined from the level set solution. Figure 9.4 illustrates that a number of particles incorrectly escape from inside the level set solution curve in the corners of the square where the characteristic information (represented by the particles themselves) needs to be deleted so that the correct vanishing viscosity solution can be obtained.

When using level set methods to model fluid flows, one is usually concerned with preserving mass (or volume for incompressible flow). Unfor-

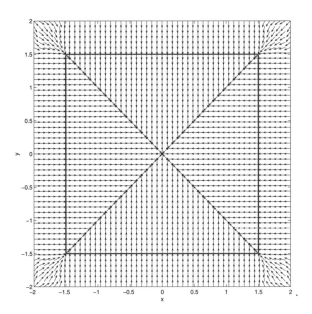

Figure 9.1. Initial square interface location and converging velocity field.

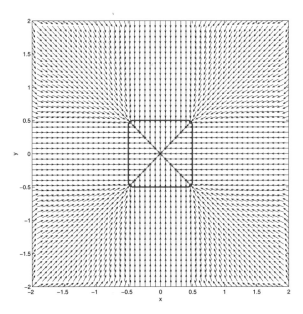

Figure 9.2. Square interface location at a later time correctly computed by the level set method.

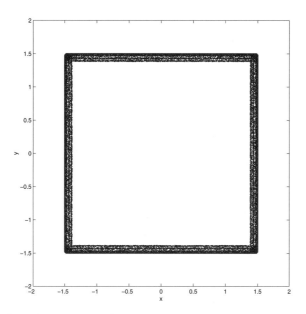

Figure 9.3. Initial square interface location and the location of a number of particles seeded interior to the interface.

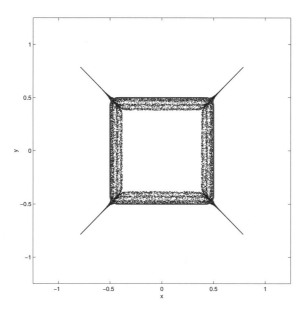

Figure 9.4. Final square interface location and the final location of the particles initially seeded interior to the interface. A number of particles have incorrectly escaped from the interior and need to be deleted in order to obtain the correct vanishing viscosity solution.

tunately, level set methods have a tendency to lose mass in underresolved regions of the flow. Attempts to improve mass (or volume) conservation in level set methods have led to a variety of reinitialization techniques, as discussed in Chapter 7. On the other hand, despite a lack of explicit enforcement of mass (or volume) conservation, Lagrangian schemes are quite successful in conserving mass, since they preserve material characteristics for all time; i.e., they are not regularized out of existence to obtain vanishing viscosity solutions. The difficulty stems from the fact that the level set method cannot accurately tell whether characteristics merge, separate, or run parallel in underresolved regions of the flow. This indeterminacy leads to vanishing viscosity solutions that can incorrectly delete characteristics when they appear to be merging.

9.2   Using Particles to Preserve Characteristics

In [61], Enright, Fedkiw, Ferziger, and Mitchell designed a hybrid particle level set method to alleviate the mass loss issues associated with the level set method. In the case of fluid flows, knowing a priori that there are no shocks present in the fluid velocity field, one can assert that characteristic information associated with that velocity field should never be deleted. They randomly seed particles near the interface and passively advect them with the flow. When marker particles cross over the interface, this indicates that characteristic information has been incorrectly deleted, and these errors are fixed by locally rebuilding the level set function using the characteristic information present in these escaped marker particles.

Since there is characteristic information on both sides of the interface, two sets of marker particles are needed. Initially, particles of both types are seeded locally on both sides of the interface, as shown in Figure 9.5. Then an equation of the form

$$\vec{x}_{\text{new}} = \vec{x} + (\phi_{\text{new}} - \phi(\vec{x}))\,\vec{N} \tag{9.1}$$

is used to attract particles initially located at $\vec{x}$ on the $\phi = \phi(\vec{x})$ isocontour to the desired $\phi = \phi_{\text{new}}$ isocontour. ϕ_{new} is chosen to place the particles on the correct side of the interface in a slightly randomized position. Figure 9.6 shows the initial placement of particles after an attraction step.

The particles are initially given a fixed radius of influence based on their distance from the interface after the seeding and attraction algorithms have been employed. As the particles are integrated forward in time using $\vec{x}_t = \vec{V}$, their position is continually monitored in order to detect possible interface crossings. When a particle crosses over the interface, indicating incorrectly deleted characteristic information, the particle's sphere of influence is used to restore this lost information. This is done with a locally applied Boolean union operation that simply adds the particle's sphere of influence to the damaged level set function; i.e., at each grid point of the

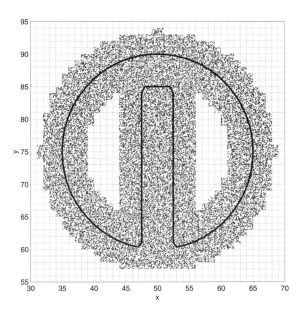

Figure 9.5. Initial placement of both types of particles on both sides of the interface. (See also color figure, Plate 1.)

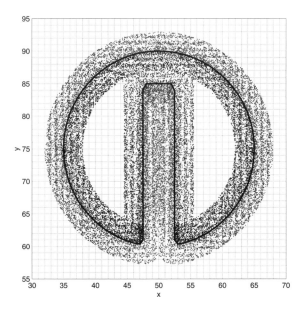

Figure 9.6. Particle positions after the initial attraction step is used to place them on the appropriate side of the interface. (See also color figure, Plate 2.)

cell containing the particle, the local value of ϕ is changed to accurately reflect the union of the particle sphere with the existing level set function.

Figures 9.7 and 9.8 show the rigid-body rotation of a notched sphere using the level set method and the particle level set method, respectively. Similarly, Figures 9.9 and 9.10 show the results of the "Enright test," where a sphere is entrained by vortices and stretched out very thin before the flow time reverses returning the sphere to its original form. The particle level set solution in Figure 9.10 returns (almost exactly) to its original spherical shape, while the level set solution in Figure 9.9 shows an 80% volume loss on the same 100^3 grid.

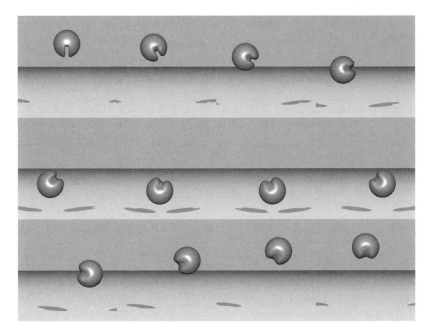

Figure 9.7. Smeared-out level set solution of a rigidly rotating notched sphere.

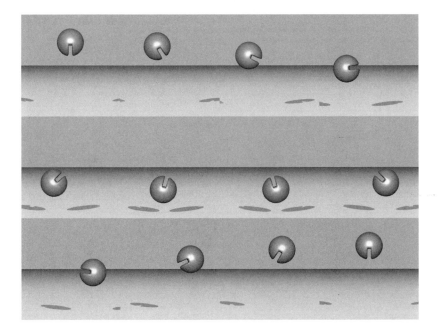

Figure 9.8. High-quality particle level set solution of a rigidly rotating notched sphere.

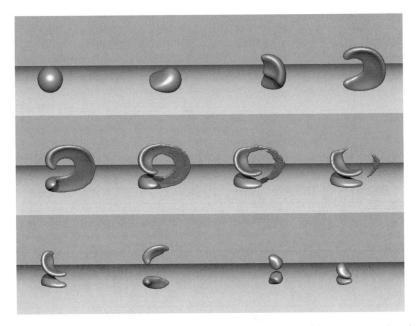

Figure 9.9. Level set solution for the "Enright test" with 80% volume loss by the final frame.

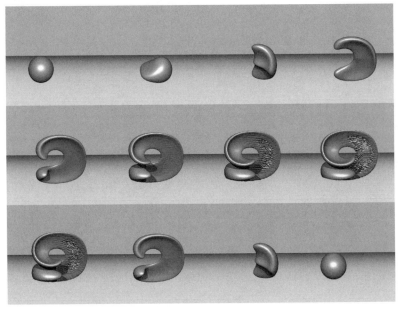

Figure 9.10. Particle level set solution for the "Enright Test." The sphere returns almost exactly to its original shape in the time reversed flow.

10
Codimension-Two Objects

10.1 Intersecting Two Level Set Functions

Typically, level set methods are used to model codimension-one objects such as points in $\Re^1$, curves in $\Re^2$, and surfaces in $\Re^3$. Burchard, Cheng, Merriman, and Osher [22] extended level set technology to treat codimension-two objects using the intersection of the zero level sets of two level set functions. That is, instead of implicitly representing codimension-one geometry by the zero isocontour of a function ϕ, codimension-two geometry is represented as the intersection of the zero isocontour of a function ϕ_1 with the zero isocontour of another function ϕ_2. In one spatial dimension, zero isocontours are points, and their intersection is usually the empty set. In two spatial dimensions, zero isocontours are curves, and the intersections of curves tend to be points which are of codimension two. In three spatial dimensions, the zero isocontours are surfaces, and the intersections of these surfaces tend to be codimension two curves.

10.2 Modeling Curves in $\Re^3$

In order to model curves as the intersection of the $\phi_1 = 0$ and $\phi_2 = 0$ isocontours of functions ϕ_1 and ϕ_2 in $\Re^3$, a number of relevant geometric quantities need to be defined. To find the tangent vectors $\vec{T}$, note that $\nabla\phi_1 \times \nabla\phi_2$, taken on the curve, is tangent to the curve. So the tangent

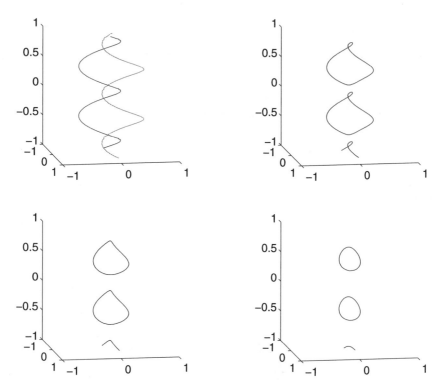

Figure 10.1. Two helices evolving under curvature motion eventually touch and merge together.

vectors are just a normalization of this:

$$\vec{T} = \frac{\nabla\phi_1 \times \nabla\phi_2}{|\nabla\phi_1 \times \nabla\phi_2|}. \tag{10.1}$$

Note that replacing ϕ_1 with $-\phi_1$ reverses the direction of the tangent vectors. This is also true when ϕ_2 is replaced with $-\phi_2$.

The curvature times the normal, $\kappa\vec{N}$, is the derivative of the tangent vector along the curve, i.e., with respect to arc length s,

$$\kappa\vec{N} = \frac{d\vec{T}}{ds}. \tag{10.2}$$

Using directional derivatives, this becomes

$$\kappa\vec{N} = \nabla\vec{T} \cdot \vec{T} = \begin{pmatrix} \nabla T_1 \cdot \vec{T} \\ \nabla T_2 \cdot \vec{T} \\ \nabla T_3 \cdot \vec{T} \end{pmatrix}, \tag{10.3}$$

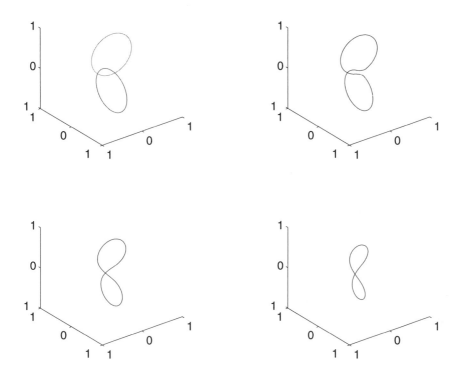

Figure 10.2. Two rings evolving under curvature motion eventually touch and merge together.

where T_1, T_2, and T_3 are the components of the tangent vector $\vec{T}$. Then the normal vectors can be defined by normalizing this quantity,

$$\vec{N} = \frac{\kappa \vec{N}}{|\kappa \vec{N}|},$$ (10.4)

and the binormal vectors are defined as

$$\vec{B} = \frac{\vec{T} \times \vec{N}}{|\vec{T} \times \vec{N}|}.$$ (10.5)

The torsion times the normal vector is defined as $\tau \vec{N} = -\nabla \vec{B} \cdot \vec{T}$. All these geometric quantities can be written in terms of ϕ_1 and ϕ_2 and computed at each grid point one uses similar to the way the normal and curvature are computed when using standard level set technology for codimension-one objects. Interpolation can be used to define these geometric quantities between grid points.

Both ϕ_1 and ϕ_2 are evolved in time using the standard level set equation. A velocity of $\vec{V} = \kappa \vec{N}$ gives curvature motion in the normal direction.

Setting $\vec{V} = \vec{N}$ or $\vec{V} = \vec{B}$ gives motion in the normal or binormal direction, respectively. Figures 10.1 and 10.2 show curves evolving under curvature motion in three spatial dimensions. In Figure 10.1 (page 88), two helices touch and merge. Similarly, in Figure 10.2 (page 89), two closed curves evolve independently until they touch and merge together.

10.3   Open Curves and Surfaces

Level set methods are used to represent closed curves and surfaces that may begin and end at the boundaries of the computational domain. However, it is not clear how to devise methods for curves and surfaces that have ends or edges (respectively) within the computational domain. Curves in $\Re^2$ have codimension two ends given by points, while surfaces in $\Re^3$ have codimension-two edges given by curves. A first step in this direction was carried out by Smereka [152] in the context of spiral crystal growth. In two spatial dimensions, he used the intersection of two level set functions ϕ and ψ to represent the codimension-two points at the beginning and end of an open curve. The curve of interest was defined as the $\phi = 0$ isocontour in the region where $\psi > 0$, while a ghost curve was defined as the $\phi = 0$ isocontour in the region where $\psi < 0$. Velocities were derived for both the curve and the fictitious ghost curve that exists only for computational purposes. Figure 10.3 shows an initial configuration where the curve moves upward and the ghost curve moves downward, as shown at a later time in Figure 10.4. Figure 10.5 shows this open curve rolling up and subsequently merging with itself, pinching off independently evolving closed curves.

10.4   Geometric Optics in a Phase-Space-Based Level Set Framework

In [124], Osher et al. introduced a level-set-based approach for ray tracing and for the construction of wave fronts in geometric optics. The approach automatically handles the multivalued solutions that appear and automatically resolves the wave fronts. The key idea, first introduced by Engquist, Runborg, and Tornberg [60], but used in a "segment projection" method rather than level set fashion, is to use the linear Liouville equation in twice as many independent variables, (actually, $2d - 1$, using a normalization) and solve in this higher-dimensional space via the idea introduced by Burchard et al. [22]. In two-dimensional ray tracing this involves solving for an evolving curve in x, y, θ space, where θ is the angle of the normal to the curve. This, of course, uses two level set functions and gives codimension-2 motion in 3-space plus time. A local level set method can be used to make

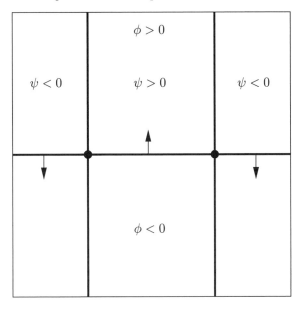

Figure 10.3. The hortizontal line marks the set where $\phi = 0$ while the two veritical lines mark the set where $\psi = 0$. The arrow in the center indicates the motion of the real curve while the arrows to the right and left indicate the motion of the ghost curves.

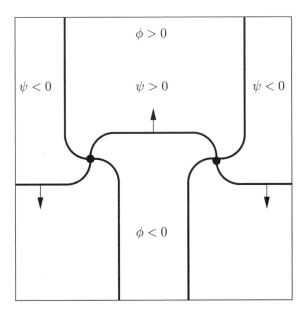

Figure 10.4. The arrow in the center indicates the motion of the real curve while the arrows to the right and left indicate the motion of the ghost curves.

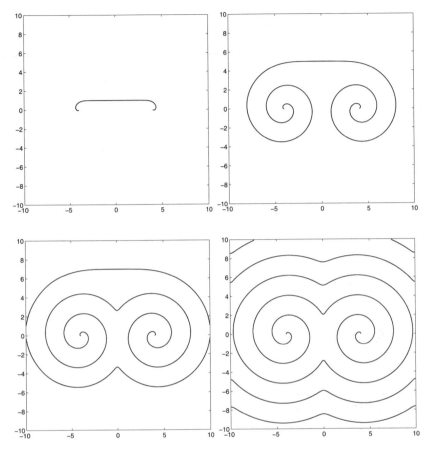

Figure 10.5. Four snapshots of the evolving open curve at various times. The curve rolls up subsequently merging with itself pinching off independently evolving closed curves.

the complexity tractable, $O(n^2 \log(n))$, for n the number of points on the curve for every time iteration. The memory requirement is $O(n^2)$.

In three-dimensional ray tracing this involves solving for an evolving two-dimensional surface in x, y, z, θ, ϕ space, where θ, ϕ give the angle of the normal, and this results in codimension-3 motion in 5-space, plus time. The complexity goes up by a power of n over the two-dimensional case, as does the memory requirement, where n is the one-dimensional number of points. Again, this involves a local level set method, this time using three level set functions.

Standard ray tracing is the ultimate Lagrangian method. Since merging and topological changes are not an issue—we actually want fronts to cross through each other without intersecting—the usual level set method has difficulties, especially with self-intersecting fronts, see, e.g., Figure 6 in [124]. However, there are many advantages of an Eulerian fixed-grid approach. Ray tracing gives inadequate spatial resolution of the wave front. This is due to the fact that points close together may diverge at later times, leaving holes in the wave front. Interpolation steps must be added. The method in [124] overcomes this resolution problem and also the usual Eulerian problem of how to get the solution when waves become multi-valued and singularities such as swallowtails or caustics develop. Eulerian approaches such as that in [59] suffer from the second problem, and the ingenious dynamic surface extension of Steinhoff et al. [156], with some improvements in Ruuth et al. [144], needs both interpolation and a method to keep track of singularities due to multiple crossing rays.

Part III
Image Processing and Computer Vision

The fields of image processing and computer vision are incredibly vast, so we do not make any attempt either to survey them or to impart any deep insight. In the following chapters we merely illustrate a few applications of level set methods in these areas.

The use of partial differential equations in image processing and computer vision, in particular the use of the level set method and dynamic implicit surfaces, has increased dramatically in recent years. Historically, the field of computer vision was probably the earliest to be affected significantly by the level set method. There are many good general references, e.g., [145, 29, 77, 120]. In this section we present three examples that are prototypes for a far wider class of applications.

The first chapter discusses a basic (perhaps the basic) issue in image processing, namely restoration of degraded images. The second concerns image segmentation with snakes and active contours. The third concerns reconstruction of surfaces from unorganized data points.

Traditionally, these closely related fields, image processing and computer vision, have developed independently. However, level set and related PDE-based methods have served to provide both a new common language and a new set of tools that have led to increased interaction.

11
Image Restoration

11.1 Introduction to PDE-Based Image Restoration

A basic idea in this field is to view a gray-scale image as a function $u_0(x, y)$ defined on a square $\Omega : \{(x, y)|0 \leq x, y \leq 1\}$, with u_0 taking on discrete values between 0 and 255, which we take as a continuum for the sake of this discussion.

A standard operation on images is to convolve u_0 with a Gaussian of variance $\sigma > 0$,

$$J(x, y, \sigma) = \frac{1}{4\pi\sigma} e^{-(x^2 + y^2)/(4\sigma)} \qquad (11.1)$$

to obtain

$$u(x, y, \sigma) = \iint J(x - x', y - y', \sigma) u_0(x', y') \, dx' \, dy' = J * u_0. \qquad (11.2)$$

This has the same effect as solving the initial value problem for the heat equation

$$\begin{aligned} u_t &= u_{xx} + u_{yy} \\ u(x, y, 0) &= u_0(x, y) \end{aligned} \qquad (11.3)$$

for $t > 0$ (ignoring boundary conditions) to obtain $u(x, y, \sigma)$ at $t = \sigma > 0$, i.e., the expression obtained in equation (11.2).

Thus, one fundamental generalization is merely to replace the heat equation with an appropriate flow equation of the form

$$u_t = F(u, Du, D^2u, x, t) \tag{11.4}$$

for $t > 0$ with intial data $u(x, y, 0) = u_0(x, y)$ defining $u(x, y, t)$ as the processed image. This is called defining a "scale space" with $t > 0$ as the scale. As t increases, the image is generally (but not always) coarsened in some sense. Here

$$Du = (u_x, u_y) \tag{11.5}$$

and

$$D^2u = \begin{pmatrix} u_{xx} & u_{xy} \\ u_{yx} & u_{yy} \end{pmatrix} \tag{11.6}$$

are the gradient and Hessian, respectively. The equation is typically of second order, as is the heat equation, although the assumption of parabolicity, especially strict parabolicity, which implies smoothing in all directions, is often weakened. Second order is usually chosen for several reasons: (1) The numerical time-step restriction is typically $\Delta t = c_1(\Delta x)^2 + c_2(\Delta y)^2$ for explicit schemes, which is reasonable. (2) The method may have a useful and appropriate maximum principle.

These generally nonlinear methods have become popular for the following reasons. Classical algorithms for image deblurring and denoising have been mainly based on least squares, Fourier series, and other L^2-norm approximations. Consequently, the results are likely to be contaminated by Gibbs's phenomenon (ringing) and/or smearing near edges. Their computational advantage comes from the fact that these classical algorithms are linear; thus fast solvers are widely available. However, the effect of the restoration is not local in space. Other bases of orthogonal functions have been introduced in order to get rid of these problems, e.g., compactly supported wavelets, but Gibbs's phenomenon and smearing are still present for these linear procedures.

Rudin [140] made the important observation that images are characterized by singularities such as edges and boundaries. Thus nonlinearity, especially ideas related to the numerical analysis of shocks in solutions of systems of hyperbolic conservation laws, should play a key role in image analysis. Later, Perona and Malik [131] described a family of nonlinear second-order equations of the type given in equation (11.4) which have an antidiffusive (hence deblurring) as well as a diffusive (hence denoising) capability. This was modified and made rigorous by Catte et al. [41] and elsewhere.

Perona and Malik proposed the following. Consider the equation

$$u_t = \nabla \cdot G(\nabla u) = \frac{\partial}{\partial x} G^1(u_x, u_y) + \frac{\partial}{\partial y} G^2(u_x, u_y) \tag{11.7}$$

with initial data $u(x, y, 0) = u(x, y)$ in Ω with appropriate boundary conditions. This equation is parabolic (hence smoothing) when the matrix of partial derivatives

$$\begin{pmatrix} G_x^1 & G_y^1 \\ G_x^2 & G_y^2 \end{pmatrix} \tag{11.8}$$

is positive definite (or weakly parabolic when it is positive semidefinite). When this matrix has a negative eigenvalue, equation (11.7) resembles the backwards heat equation. One might expect such initial value problems to result in unstable blowup, especially for nonsmooth initial data. However, if we multiply equation (11.7) by $u^{2p-1}/(2p)$ for p a positive integer and integrate by parts, we arrive at

$$\frac{d}{dt} \int_\Omega |u|^{2p} \, d\Omega = - \int \left(\frac{2p-1}{2p} \right) (\nabla u \cdot G(\nabla u)) \, |u|^{2p-2} \, d\Omega. \tag{11.9}$$

So if

$$u_x G^1(u_x, u_y) + u_y G^2(u_x, u_y) \geq 0, \tag{11.10}$$

the solutions stay bounded in all L^p, $p > 1$, spaces, including $p = \infty$, which means that the maximum of u is nonincreasing. Then all one needs to do is allow G to be backwards parabolic but satisfy equation (11.10) above, and a restoring effect is obtained.

This initial value problem is not well posed; i.e., two nearby initial data will generally result in very different solutions. There are many such examples in the literature. In one spatial dimension parabolicity means that the derivative of $G^1(u_x)$ is positive, while equation (11.10) just means that $u_x G^1(u_x) \geq 0$. Obviously, functions that have the same sign as their argument but are sometimes decreasing in their argument will give bounded restoring results in those ranges of u_x.

11.2   Total Variation-Based Image Restoration

The total variation of a function of one variable $u(x)$ can be defined as

$$TV(u) = \sup_{h>0} \int \left| \frac{u(x+h) - u(x)}{h} \right| \, dx, \tag{11.11}$$

which we will pretend is the same as $\int |u_x| dx$ (analogous statements are made in two or more spatial dimensions). Thus $TV(u)$ is finite for any bounded increasing or decreasing function, including functions with jump discontinuities. On the other hand, this is not true for $\int |u_x|^p dx$ for any $p > 1$. Note that $p < 1$ results in a nonconvex functional; i.e., the triangle inequality is false. For this reason (among others) TV functions are the appropriate class for shock solutions to conservation laws and for the problems arising in processing images with edges.

We also remark that it is very easy to show (formally at least) that $u_x G(u_x) \geq 0$ implies that the evolution

$$u_t = \frac{\partial}{\partial x} G(u_x) \tag{11.12}$$

leads to the estimate

$$\frac{d}{dt} \int |u_x| \, dx \leq 0, \tag{11.13}$$

i.e., the evolution is total variation diminishing or TVD; see e.g., Harten [80]. To see this, multiply equation (11.12) by sign u_x and proceed formally. In fact it is easy to see that the finite difference approximation to equation (11.12),

$$u_i^{n+1} = u_i^n + \frac{\Delta t}{\Delta x} \left(G\left(\frac{u_{i+1}^n - u_i^n}{\Delta x} \right) - G\left(\frac{u_i^n - u_{i-1}^n}{\Delta x} \right) \right), \tag{11.14}$$

with a time-step restriction of

$$\frac{\Delta t}{(\Delta x)^2} \left(\frac{G(u_x)}{u_x} \right) < \frac{1}{2} \tag{11.15}$$

satisfies

$$\sum_i |u_{i+1}^{n+1} - u_i^{n+1}| \leq \sum_i |u_{i+1}^n - u_i^n|; \tag{11.16}$$

i.e., it is also TVD.

All this makes one realize that TV, as predicted by Rudin [140], is in some sense the right class in which to process images, at least away from highly oscillatory textures. The work of Rudin [140] led to the TV preserving shock filters of Rudin and Osher [125] and to the successful total variation based restoration algorithms of Rudin et al. [142, 141] and Marquina and Osher [111].

In brief, suppose we are presented with a noisy blurred image

$$u_0 = J * u + n, \tag{11.17}$$

where J is a given convolution kernel (see equation (11.2)) and n represents noise. Suppose also that we have some estimate on the mean and variance of the noise. We then wish to obtain the "best" restored image. Our definition of "best" must include an accurate treatment of edges, i.e., jumps.

A straightforward approach is just to invert this directly. If the Fourier transform of J, $\hat{J} = FJ$, is nonvanishing, then we could try

$$u = F^{-1}(\hat{J}^{-1}\hat{u}_0). \tag{11.18}$$

This gives poor results, even in the absence of noise, because high frequencies associated with edges will be amplified, leading to major spurious oscillations. Of course, the presence of significant noise makes this procedure disastrous.

Another approach might be to regularize this procedure. We could try adding a penalty term to a minimization procedure; i.e., choose u to solve the following constrained minimization problem:

$$\min_u \left(\mu \int |\nabla u|^2 d\Omega + \|J * u - u_0\|_{L^2}^2 \right), \qquad (11.19)$$

where $\hat{\lambda} = \mu^{-1} > 0$ is often a Lagrange multiplier. The minimizer is

$$u = F^{-1} \left(\frac{\hat{J}(\xi_1, \xi_2)\hat{u}_0(\xi_1, \xi_2)}{\mu(\xi_1^2 + \xi_2^2) + |\hat{J}(\xi_1, \xi_2)|^2} \right), \qquad (11.20)$$

where μ can be chosen so that the variance of the noise is given; i.e., we can choose μ so that

$$\int |J * u - u_0|^2 d\Omega = \sigma^2 = \int |\hat{J}\hat{u} - \hat{u}_0|^2 d\Omega, \qquad (11.21)$$

which is a simple algebraic equation for $\hat{\lambda}$. Alternatively, μ could be a coefficient in a penalty term and could be obtained by trial and error. Obviously, this is a very simple and fast procedure, but the effect of it is either to smear edges or to allow oscillations. This is because the space of functions we are considering does not allow clean jumps.

Instead, the very simple idea introduced by Rudin et al. [142] is merely to replace the power of 2 by a 1 in the exponent of $|\nabla u|$ in equation (11.19). Thus, TV restoration is

$$\min_u \left(\int |\nabla u| \, d\Omega + \hat{\lambda} \|J * u - u_0\|^2 \right), \qquad (11.22)$$

where $\hat{\lambda} > 0$ is a Langrange multiplier and we drop the L_2 subscript in the second term. This leads us to the nonlinear Euler-Lagrange equations (assuming $J(x) = J(-x)$ for simplicity only)

$$\nabla \cdot \left(\frac{\nabla u}{|\nabla u|} \right) - \lambda J * (J * u - u_0) = 0, \qquad (11.23)$$

where $\lambda = 2\hat{\lambda}$. Of course, Fourier analysis is useless here, so the standard method of solving this is to use gradient descent, i.e., to solve

$$u_t = \nabla \cdot \left(\frac{\nabla u}{|\nabla u|} \right) - \lambda J * (J * u - u_0) \qquad (11.24)$$

for $t > 0$ to steady state with $u(x, y, 0)$ given. Again λ may be a chosen so as to satisfy equation (11.21), although the procedure to enforce this is a bit more intricate. First we note that if the mean of the noise is zero, i.e., $\int n \, d\Omega = 0$ and $\int J d\Omega = 1$, it is easy to see that the constraint

$$\int (J * u(x, y, t) - u_0) \, d\Omega = 0 \qquad (11.25)$$

is satisfied for all $t > 0$ if it is satisfied at $t = 0$. This is true regardless of the choice of λ.

In order to satisfy the second constraint, equation (11.21), we use a version of the projection gradient method of Rosen [138] introduced in [142]. We note that a nonvariational version of this idea was used by Sussman and Fatemi [158] to help preserve area for the level set reinitialization step; see Chapter 7. If we wish to solve $\min_u \int f(u) \, d\Omega$ such that $\int g(u) \, d\Omega = 0$, we start with a function V_0 such that $\int g(V_0) \, d\Omega = 0$. Then gradient descent leads us to the evolution equation

$$u_t = -f_u - \lambda g_u \tag{11.26}$$

for $t > 0$ with $u(0) = V_0$. We wish to maintain the constraint under the flow, which means

$$\frac{d}{dt} \int g(u) \, d\Omega = 0 = \int g_u u_t \, d\Omega = -\int f_u g_u \, d\Omega - \lambda \int g_u^2 d\Omega; \tag{11.27}$$

thus we merely choose $\lambda(t)$ such that

$$\lambda(t) = -\frac{\int f_u g_u \, d\Omega}{\int g_u^2 \, d\Omega}. \tag{11.28}$$

Thus we have

$$u_t = -f_u + g_u \left(\frac{\int f_u g_u \, d\Omega}{\int g_u^2 \, d\Omega} \right) \tag{11.29}$$

and

$$\frac{d}{dt} \int f(u) \, d\Omega = -\int f_u^2 \, d\Omega + \frac{\left(\int f_u g_u \, d\Omega \right)^2}{\int g_u^2 \, d\Omega} \leq 0, \tag{11.30}$$

by Schwartz's inequality, so the function to be minimized diminishes (strictly, if g_u and f_u are independent), and convergence occurs when u is such that $f_u + \lambda g_u = 0$ for some λ. These ideas generalize to much more complicated situations, e.g., many independent constraints. In our present setting this leads us to choosing λ in equation (11.24) as

$$\lambda = \frac{\int \nabla \cdot \left(\frac{\nabla u}{|\nabla u|} \right) (J * (J * u - u_0)) \, d\Omega}{\|J * (J * u - u_0)\|^2}. \tag{11.31}$$

Thus we have our TV denoising/deblurring algorithm given by equations (11.24) and (11.31). Again, we repeat that in practice μ is often picked by the user to be a fixed constant. Another difficulty with using equation (11.31) comes in the initialization. Recall that we need $u(x, y, 0)$ to satisfy equation (11.25) as well as

$$(J * u(x, y, 0) - u_0)^2 = \sigma^2. \tag{11.32}$$

11.3 Numerical Implementation of TV Restoration

We now turn to the issue of fast numerical implementation of this method, as well as its connection with the dynamic evolution of surfaces. The evolution equation (11.24) has an interesting geometric interpretation in terms of level set evolution. We can view this as a procedure that first moves every level set of the function u with velocity equal to its mean curvature divided by the norm of the gradient of u and then projects back onto the constraint set, for which the variance of the noise is fixed. The first step has the effect of removing high-curvature specks of noise, even in the presence of steep gradients, and leaving alone piecewise smooth clean functions at their jump discontinuities.

From a finite difference scheme point of view, the effect at edges is easy to describe. Suppose we wish to approximate equation (11.24) for $\lambda = 0$. The equation can be written as

$$u_t = \frac{\partial}{\partial x}\left(\frac{u_x}{\sqrt{u_x^2 + u_y^2 + \delta}}\right) + \frac{\partial}{\partial y}\left(\frac{u_y}{\sqrt{u_x^2 + u_y^2 + \delta}}\right), \qquad (11.33)$$

where $\delta > 0$ is very small, chosen to avoid division by zero at places where $|\nabla u| = 0$. There are serious numerical issues here involving time-step restrictions. Intuitively, an explicit scheme should be restricted by

$$\frac{\Delta t}{(\Delta x)^2} \le c\sqrt{u_x^2 + u_y^2 + \delta} \qquad (11.34)$$

for a constant c, which is terribly restrictive near flat (zero) gradients.

A typical scheme might be

$$u_{i,j}^{n+1} = u_{i,j}^n - \frac{\Delta t}{(\Delta x)^2}\left(C_{i+\frac{1}{2},j}^n(u_{i+1,j}^n - u_{i,j}^n) - C_{i-\frac{1}{2},j}^n(u_{i,j}^n - u_{i-1,j}^n)\right.$$
$$\left. + D_{i,j+\frac{1}{2}}^n(u_{i,j+1}^n - u_{i,j}^n) - D_{i,j-\frac{1}{2}}^n(u_{i,j}^n - u_{i,j-1}^n)\right), \qquad (11.35)$$

where

$$C_{i+\frac{1}{2},j} = \left((\Delta_+^x u_{i,j})^2 + \frac{(\Delta_+^y u_{i+\frac{1}{2},j})^2}{2} + \frac{(\Delta_-^y u_{i+\frac{1}{2},j})^2}{2} + \delta(\Delta x)^2\right)^{-\frac{1}{2}} > 0$$
$$(11.36)$$

and

$$D_{i,j+\frac{1}{2}} = \left((\Delta_+^y u_{i,j})^2 + \frac{(\Delta_+^x u_{i,j+\frac{1}{2}})^2}{2} + \frac{(\Delta_-^x u_{i,j+\frac{1}{2}})^2}{2} + \delta(\Delta x)^2\right)^{-\frac{1}{2}} > 0$$
$$(11.37)$$

for

$$u_{i+\frac{1}{2},j} = \frac{1}{2}u_{i,j} + \frac{1}{2}u_{i+1,j},$$

$$u_{i,j+\frac{1}{2}} = \frac{1}{2}u_{i,j} + \frac{1}{2}u_{i,j+1}.$$

(11.38)

This is a second-order accurate spatial discretization.

At steady state $u_{i,j}^{n+1} \equiv u_{i,j}^{n} \equiv u_{i,j}$, and we can solve for $u_{i,j}^{n}$, obtaining the nonlinear relationship

$$u_{i,j} = \frac{C_{i+\frac{1}{2},j}u_{i+1,j} + C_{i-\frac{1}{2},j}u_{i-1,j} + D_{i,j+\frac{1}{2}}u_{i,j+1} + D_{i,j-\frac{1}{2}}u_{i,j-1}}{C_{i+\frac{1}{2},j} + C_{i-\frac{1}{2},j} + D_{i,j+\frac{1}{2}} + D_{i,j-\frac{1}{2}}},$$

(11.39)

i.e., $u_{i,j}$ is a weighted convex combination of its four neighbors. This smoothing is anisotropic. If, for example, there is a large jump from $u_{i,j}$ to $u_{i+1,j}$, then $C_{i+1/2,j}$ is close to zero, and the weighting is done in an almost WENO fashion. This helps to explain the virtues of TV denoising: The edges are hardly smeared. Contrast this with the linear heat equation, which at steady state yields

$$C_{i,j} = \frac{1}{4}[u_{i+i,j} + u_{i-1,j} + u_{i,j+1} + u_{i,j-1}].$$

(11.40)

Severe smearing of edges is the result. For further discussion and generalizations, see Chan et al. [31].

The explicit time-step restriction in equation (11.34) leads us to believe that convergence to steady state might be slow, especially in the presence of significant blur and noise. Many attempts have been made to speed this up; see e.g., Chan et al. [30]. Another interesting observation is that equation (11.24) scales in a strange way. If u is replaced by $h(u)$, with $h' > 0$, $h(0) = 0$, and $h(255) = 255$, then equation (11.24) even for $\lambda = 0$ is not invariant. This means that the evolution process is not morphological; see Alvarez et al. [5]; i.e., it does not depend only on the level sets of the intensity, but on their values. One possible fix for these two problems is the following simple idea introduced by Marquina and Osher [111].

We merely multiply the right-hand side of equation (11.24) by $|\nabla u|$ and drive this equation to steady state. The effect of this is beneficial in various aspects. Although the steady states of both models are analytically the same, since $|\nabla u|$ vanishes only in flat regions, there are strong numerical, analytical, and philosophical advantages of this newer model.

(1) The time-step restriction is now $\Delta t/(\Delta x)^2 \le c$ for some $c > 0$, so simple explicit-in-time methods can be used.

(2) We can use ENO or WENO versions of Roe's entropy-condition-violating scheme for the convection term (there is no viscosity or entropy condition for images) and central differencing for the regularized anisotropic diffusion (curvature) term. This seems to give

better numerical answers; the numerical steady states do not have the staircasing effect sometimes seen in TV reconstruction; see, e.g., Chambolle and Lions [42].

(3) In the pure denoising case, i.e., $J * u \equiv u$, there is a simple maximum principle (analytical as well as numerical).

(4) The procedure is almost morphological; i.e., if we replace u by $h(v)$ and u_0 by $h(v_0)$ with $h'(u) > 0$, then the evolution is transformed as follows:

$$u_t = |\nabla u| \, \nabla \cdot \left(\frac{\nabla u}{|\nabla u|} \right) - \lambda |\nabla u| J * (J * u - u_0) \qquad (11.41)$$

transforms to

$$v_t = |\nabla v| \nabla \cdot \left(\frac{\nabla v}{|\nabla v|} \right) - \lambda |\nabla v| J * (J * h(v) - h(v_0)) \qquad (11.42)$$

i.e., we still have motion by mean curvature followed by the slightly modified projection on the constraint set.

The maximum principle is a mixed blessing in this case. If $J * u = u$, i.e., we are doing the pure denoising case, then the fact that extrema are not amplified is a good thing, and we may take as the initial data $u(x, y, 0) = u_0(x, y)$; i.e., the noisy image is a good initial guess. This is not a good choice for the deblurring case. There we merely use the linear deconvolved approximation in equation (11.20) where μ is chosen to match the constraint equation (11.21). Although this introduces spurious oscillations in the initial guess, they seem to disappear rapidly when equation (11.41) is used.

An interesting feature of this new approach is that we can view the right-hand side of equation (11.41) as consisting of an elliptic term added to a Hamilton-Jacobi term. The elliptic term uses standard central differences, while the Hamilton-Jacobi term is upwinded according to the direction of characteristics. What is a bit unusual here is that there should be no entropy fix in the approximate Hamiltonian. The viscosity criterion does not apply, and Roe's entropy-violating scheme is used. For details see [111].

To repeat, one consequence of this approach is that although the steady states of equations (11.24) and (11.41) are the same, the numerical solutions differ, and staircasing, as described in [42], seems to be minimized using equation (11.41).

We demonstrate the results of our improved algorithms with the following experiments. Figure 11.1 shows a noisy piecewise linear one-dimensional signal with a signal-to-noise ratio approximately equal to 3. Figure 11.2 shows the recovered signal wtih denoised edges, but also with staircasing effects based on the original TV method developed in [142]. Figure 11.3 shows the improved result without staircase effects in the linear region; see [111].

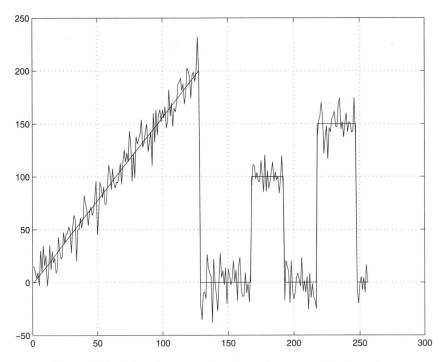

Figure 11.1. Original versus noisy signal in one spatial dimension.

Figures 11.4 and 11.5 show an original image and a noisy (signal-to-noise ratio approximately 3) image, respectively. Figure 11.6 shows the usual TV recovered image, while Figure 11.7 uses the method in [111] and seems to do better in recovering smooth regions.

Figure 11.8 represents an image blurred by a discrete Gaussian blur obtained by solving the heat equation with Figure 11.4 as initial data and computing the solution at $t = 10$ on a 128×128 grid. Figure 11.9 shows the result of an approximate linear deconvolution. Note that in the absence of noise the result is oscillatory but greatly improved. Figure 11.10 shows the result of using Figure 11.9 as an initial guess for our improved TV restoration. Notice the good resolution without spurious oscillations.

Our most demanding experiment was performed on the blurry and noisy image obtained from the original image represented in Figure 11.11. The experimental point-spread function $j(x, y)$ is shown in Figure 11.12, and we add Gaussian noise, signal-to-noise ratio approximately 5. The blurry noisy image is shown in Figure 11.13. The linear recovery is shown in Figure 11.14. Finally, the improved TV restoration using Figure 11.14 as initial guess is shown in Figure 11.15.

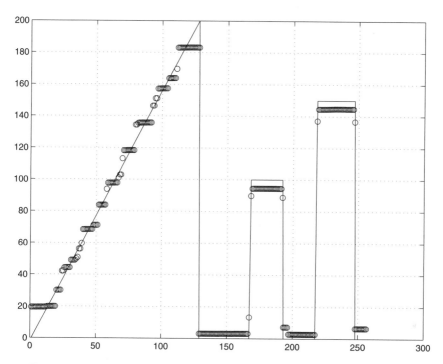

Figure 11.2. TV recovery using the original method of Rudin et al. [142].

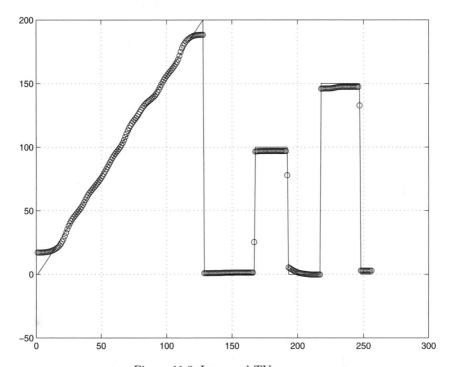

Figure 11.3. Improved TV recovery.

Figure 11.4. Original image.

Figure 11.5. Noisy image, signal-to-noise ratio approximately equal to 3.

Figure 11.6. Usual TV recovered image.

Figure 11.7. Improved TV recovered image.

Figure 11.8. Image blurred by convolution with a Gaussian kernel.

Figure 11.9. Linear deconvolution applied to Figure 11.8.

Figure 11.10. Improved TV restoration of Figure 11.8 using Figure 11.9 as an initial guess.

Figure 11.11. Original image.

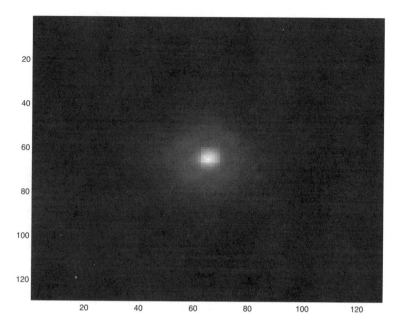

Figure 11.12. Experimental point-spread function.

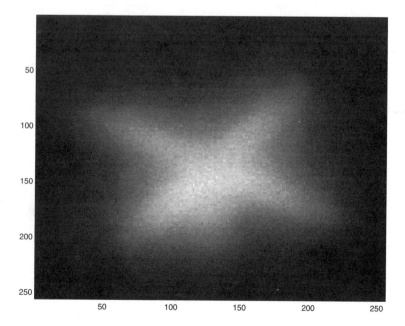

Figure 11.13. Blurry noisy version of Figure 11.11.

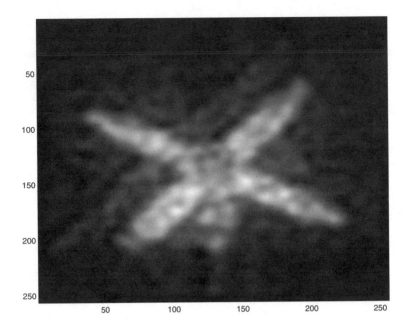

Figure 11.14. Linear restoration of Figure 11.13.

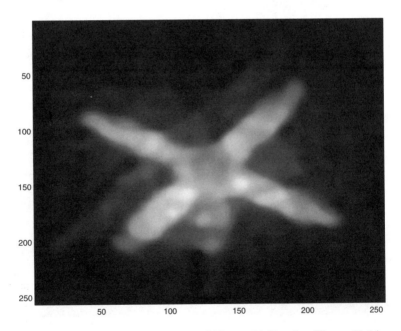

Figure 11.15. Improved TV restoration of Figure 11.13 using Figure 11.14 as an initial guess.

12
Snakes, Active Contours, and Segmentation

12.1 Introduction and Classical Active Contours

The basic idea in active contour models (or snakes) is to evolve a curve, subject to constraints from a given image u_0, in order to detect objects in that image. Ideally, we begin with a curve around the object to be detected, and the curve then moves normal to itself and stops at the boundary of the object. Since its invention by Kass et al. [94] this technique has been used both often and successfully. The classical snakes model in [94] involves an edge detector, which depends on the gradient of the image u_0, to stop the evolving curve at the boundary of the object.

Let $u_0(x, y)$ map the square $0 \leq x, y \leq 1$ into R, where u_0 is the image and $C(I) : [0, 1] \to R^2$ is the parametrized curve. The snake model is to minimize

$$F_1(C) = \alpha \int_0^1 |C'(s)|^2 \, ds + \beta \int_0^1 |C''(s)| \, ds - \lambda \int_0^1 |\nabla u_0(C(s))|^2 \, ds, \quad (12.1)$$

where α, β, and λ are positive parameters. The first two terms control the smoothness of the contour, while the third attracts the contour toward the object in the image (the external energy). Observe that by minimizing the energy, we are trying to locate the curve at the points of maximum $|\nabla u_0|$, which act as an edge detector, while keeping the curve smooth.

An edge detector can be defined by a positive decreasing function $g(\vec{z})$, depending on the gradient of the image u_0, such that

$$\lim_{|\vec{z}| \to \infty} g(\vec{z}) = 0.$$

A typical example is

$$g(\nabla u_0(\vec{x})) = \frac{1}{1 + |J * \nabla u_0|^p}$$

for $p \geq 1$, where J is a Gaussian of variance σ.

Rather than using the energy defined in equation (12.1), we can define a compact version as in Caselles et al. [28] or Kichenassamy et al. [95] via

$$F_2(C) = \int_0^1 g(\nabla(u_0(s))) \, ds. \tag{12.2}$$

Using the variational level set formulation of Zhao et al. [175], we arrive at

$$\phi_t = |\nabla\phi|\nabla \cdot \left[g(\nabla u_0) \left(\frac{\nabla\phi}{|\nabla\phi|} \right) \right] \tag{12.3}$$

$$= |\nabla\phi| \left(g(\nabla u_0)\kappa + \nabla g(\nabla u_0) \cdot \frac{\nabla\phi}{|\nabla\phi|} \right)$$

$$= |\nabla\phi| g(\nabla u_0)\kappa + \nabla g(\nabla u_0) \cdot \nabla\phi.$$

This is motion of the curve with normal velocity equal to its curvature times the edge detector plus convection in the direction that is the gradient of the edge detector. Thus, the image gradient determines the location of the snakes.

The level set formulation of this came after the original snake model was invented in [94]. This was first done in Caselles et al. [27] (without the convection term), next by Malladi et al. [109], and the variational formulation used above came in [28] and [95]. Of course, this level set formulation allows topological changes and geometrical flexibility, and has been quite successful in two and three spatial dimensions. Most models have the same general form as equation (12.3), involving an edge detector times curvature plus linear advection. The higher-order terms coming from the term multiplying β in equations (12.1) and (12.2) are usually omitted. We note that this model depends on the image gradient to stop the curve (or surface) evolution.

In a sequence of papers beginning with Chan and Vese [35] (see also [34] and [32]) the authors propose a different active contour model without a stopping i.e. edge function, i.e., a model that does not use the gradient of the image u_0 for the stopping process. The stopping term is based on the Mumford-Shah segmentation technique, which we describe below. The model these authors develop can detect contours both with and without gradients, for instance objects that are very smooth, or even have discontinuous boundaries. In addition, the model and its level set formulation are such that interior contours are automatically detected, and the initial curve can be anywhere in the image.

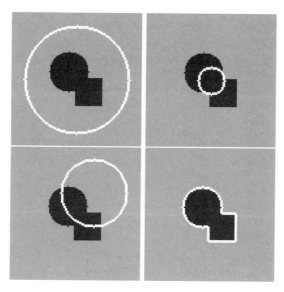

Figure 12.1. A simple case, showing that the fitting term $E_1(\Gamma)$ is minimized when the curve is on the boundary of the object.

12.2 Active Contours Without Edges

Define the evolving curve Γ as the boundary of a region Ω. We call Ω the inside of Γ and the complement of $\Omega = \Omega^c$ the outside of Γ. The method is the minimization of an energy-based segmentation. Assume that u_0 is formed by two regions of approximately piecewise constant intensities of distinct values u_0^i and u_0^o. Assume further that the object to be detected is represented by the region with value u_0^i. Denote its boundary by Γ_0. Then we have $u_0 \approx u_0^i$ inside Γ_0 and $u_0 \approx u_0^o$ outside Γ_0. Now consider the "fitting" term

$$E_1(\Gamma) = \int_{\text{inside } \Gamma} |u_0(\vec{x}) - C_1|^2 \, d\vec{x} + \int_{\text{outside } \Gamma} |u_0(\vec{x}) - C_2|^2 \, d\vec{x}, \quad (12.4)$$

where Γ is any curve and C_1, C_2 are the averages of u_0 inside Γ and outside Γ. In this simple case it is obvious that Γ_0, the boundary of the object, is the minimizer of the fitting term. See Figure 12.1.

In the active contour model proposed in [35] and [34] the fitting term plus some regularizing terms will be minimized. The regularizing terms will involve the length of the boundary Γ and the area of Ω, the region inside Γ. This is in the spirit of the Mumford-Shah functional [117]. Thus, using the variational level set formulation [175], the energy can be written, with ϕ

the level set function associated with Ω, as

$$E(C_1, C_2, \phi) = \mu \int \delta(\phi) |\nabla \phi| \, d\vec{x} \tag{12.5}$$

$$+ \nu \int H(\phi) \, d\vec{x}$$

$$+ \lambda_1 \int |u_0(\vec{x}) - C_1|^2 H(\phi) \, d\vec{x}$$

$$+ \lambda_2 \int |u_0(\vec{x}) - C_2|^2 (1 - H(\phi)) \, d\vec{x}.$$

This involves the four nonnegative parameters μ, ν, λ_1, and λ_2.

The classical Mumford-Shah functional is a more general segmentation defined by

$$E_{MS}(\Gamma, u) = \mu \, \text{length}(\Gamma) \tag{12.6}$$

$$+ \lambda \int |u - u_0|^2 \, d\vec{x}$$

$$+ \nu \int_{\Gamma^c} |\nabla u|^2 \, d\vec{x}.$$

Here u is the cartoon image approximating u_0, u is smooth except for jumps on the set Γ of boundary curves, and Γ segments the image into piecewise smooth regions. The method defined in equation (12.5) differs from that in equation (12.6) in that only two subregions are allowed in which u is piecewise constant, so we may write

$$u = C_1 H(\phi) + C_2(1 - H(\phi)). \tag{12.7}$$

This was generalized considerably by Chan and Vese [36, 37]. We also mention the approach of Koepfler et al. [99], which approximates equation (12.6) by letting $\nu = 0$ and $\lambda = 1$, where μ is the scale parameter. Again, u is piecewise constant, although many constants are allowed; i.e., the averages of u will generally be different in different segments of the image. The parameter μ defines the scale for the method in the sense that if $\mu = \infty$, then the length of the boundaries should be minimized. So we take u to be the average of u_0 over the whole image; this is the coarsest scale. If $\mu = 0$, then there is no penalty for length; each grid point (pixel) is the average of u_0, or just the value of u_0; and the segmentation u is equal to the original image. As μ increases the segmentation coarsens. This is the idea behind the segmentation of [99]; it is a split and merge method, not a partial differential equations-based approach, as in [36, 37].

Returning to the model in equation (12.5), it is easy to see that with respect to the constants C_1 and C_2 it is easy to express these two in terms of ϕ as

$$C_1(\phi) = \frac{\int u_0(\vec{x}) H(\phi(\vec{x})) \, d\vec{x}}{\int H(\phi(\vec{x})) \, d\vec{x}}, \tag{12.8}$$

$$C_2(\phi) = \frac{\int u_0(\vec{x})(1 - H(\phi(\vec{x}))) \, d\vec{x}}{\int (1 - H(\phi(\vec{x}))) \, d\vec{x}}. \tag{12.9}$$

This expresses the fact that the best constant value for the segment u is just the average of u_0 over the subregion.

In order to compute the Euler-Lagrange equations we use the variational level set approach and arrive at

$$\frac{\partial \phi}{\partial t} = |\nabla \phi| \left[\mu \nabla \cdot \left(\frac{\nabla \phi}{|\nabla \phi|} \right) - \nu - \lambda_1 (u_0 - C_1)^2 + \lambda_2 (u_0 - C_2)^2 \right]$$

$$\tag{12.10}$$

$$\phi(\vec{x}, 0) = \phi_0(\vec{x}). \tag{12.11}$$

However, it was found in [35, 34], that the nonmorphological approach was more effective; i.e., $|\nabla \phi|$ is replaced by $\delta_\epsilon(\phi)$ in the term multiplying the brackets in equation (12.10). Here

$$\delta_\epsilon(z) = \frac{\partial}{\partial z} H_\epsilon(z) = \frac{\partial}{\partial z} \frac{1}{2} \left(1 + \frac{2}{\pi} \tan^{-1} \left(\frac{z}{\epsilon} \right) \right) \tag{12.12}$$

for $\epsilon > 0$ and small, which gives a globally positive approximation to the delta function. This is necessary, as we shall discuss below. Thus the model defined in [35, 34] is

$$\frac{\partial \phi}{\partial t} = \delta_\epsilon(\phi) \left[\mu \nabla \cdot \left(\frac{\nabla \phi}{|\nabla \phi|} \right) - \nu - \lambda_1 (u_0 - C_1)^2 + \lambda_2 (u_0 - C_2)^2 \right] \tag{12.13}$$

with C_1 and C_2 defined in equations (12.8) and (12.9). Generally, the parameters are taken to be $\nu = 0$, $\lambda_1 = \lambda_2 = 1$, and $\mu > 0$ is the scale parameter. Although only two regions Ω and Ω^c can be constructed, they can, and generally will, be disconnected into numerous components in the fine-scale case, with each component having one of two constant values for u.

One important remark concerning this model as opposed to other level set evolutions is its global nature. All level sets of ϕ have the potential to be important. This means that other isocontours corresponding to nonzero values of ϕ might evolve so as to push through the $\phi = 0$ barrier and create new segmented regions. Thus reinitialization to the distance function is not a good idea here (as pointed out by Fedkiw). One can even begin with $\phi > 0$ or $\phi < 0$ throughout the region and watch new zeros develop. Of course, this also explains why we need $\delta_\epsilon(z) > 0$ in equation (12.12). Again, the goal here is to detect interior contours. The technical reason why this works is that the image u_0 acts in a nontrivial and nonlinear way as a forcing function on all the level contours of ϕ, forcing some to go through 0 spontaneously.

The scale parameter should be small if we have to detect many objects, including small objects. If we have to detect only large objects (for example a cluster of lights), the scale parameter μ should be larger. An extremely trivial but slightly instructive analytic example is the following. Let $\mu = 0$ and $\lambda_1 = \lambda_2 = \lambda$, so the finest-scale segmentation occurs; i.e., every point should be a boundary point and $u(\vec{x}) = u_0(\vec{x})$. That this does occur follows from the evolution

$$\frac{\partial \phi}{\partial t} = \delta_\epsilon(\phi) \left[\lambda(C_2 - C_1) \left[u_0 - \left(\frac{C_1 + C_2}{2} \right) \right] \right], \qquad (12.14)$$

so steady state can occur only if the average of u_0 over Ω equals its average over Ω^c (which is an unstable equilibrium) or if $\phi \equiv 0$, which is the desired equilibrium. If we take $\mu > 0$ and take the limit as $\mu \to 0$, we believe intuitively that the $u = u_0$, $\phi \equiv 0$ solution is the stable limit, since the curvature term will tend to move the boundaries when these averages happen to be equal. Numerical experiments indicate that this is true, and hence an infinite (actually as many as there are grid points) number of new zero-level contours develop in a stable fashion.

12.3 Results

We illustrate in Figures 12.2, 12.3, 12.4 and 12.5 the main advantages of this active contour model without edges [34]: detection of cognitive contours (which are not defined by gradients) in Figure 12.2, detection of contours in a noisy image in Figure 12.3, detection of interior contours automatically and extension to three dimensions in Figures 12.4 and 12.5. Also, note that the initial curve does not need to enclose the objects, as in the classical snakes and as in active contour models based on the gradient-edge detector.

12.4 Extensions

As in the original Mumford-Shah functional [117] and the implementations of [99] and [34], one may propose the use of other channels, e.g., replacing u_0 by the curvature of its level sets $\nabla \cdot (\nabla u_0/|\nabla u_0|)$, or by their orientations $u_0 = \tan^{-1}((u_0)_x/(u_0)_y)$, to do texture segmentation.

Chan et al. [32] extended the method to vector-valued images as follows. Let $u_{0,i}$ be the ith channel of an image on the usual square region with N channels and Γ the evolving curve. See [32] for examples of these channels, which include color images. The extension to the vector case is straightforward. The level set evolution becomes

$$\frac{\partial \phi}{\partial \epsilon} = \delta_\epsilon(\phi)\Big[\mu \nabla \cdot \left(\frac{\nabla \phi}{|\nabla \phi|}\right) - \frac{1}{N}\sum_{i=1}^{N} \lambda_i^{+}(u_{0,i} - C_i^{+})^2 \tag{12.15}$$

$$+ \frac{1}{N}\sum_{i=1}^{N} \lambda_i^{-}(u_{0,i} - C_i^{-})^2\Big].$$

Here the $C_i^{\pm}$ are the averages of $u_{0,i}$ on $\phi > 0$ and $\phi < 0$, respectively.

Another extension, again using only one level set function, involves removing the piecewise constant assumption and allowing piecewise-smooth solutions to the variational problem, smooth inside each zero isocontour of ϕ, with jumps across the edges, as described in Chan and Vese [37]. This is in the spirit of the original Mumford-Shah functional, although multiple junctions are not yet allowed; see below for that. The minimization procedure is

$$\inf_{u^+, u^-, \phi} F(u^+, u^- \phi), \tag{12.16}$$

where

$$F(u^+, u^-, \phi) = \int |u^+ - u_0|^2 H(\phi) d\vec{x} \tag{12.17}$$

$$+ \int |u^- - u_0|^2 (1 - H(\phi)) d\vec{x}$$

$$+ \nu \int |\nabla u^+|^2 H(\phi) d\vec{x}$$

$$+ \nu \int |\nabla u^-|^2 (1 - H(\phi)) d\vec{x}$$

$$+ \mu \int |\nabla H(\phi)| d\vec{x}.$$

The Euler-Lagrange equations for u^+ and u^- are

$$u^+ - u_0 = \nu \Delta u^+ \text{ on } \{\vec{x} \mid \phi(\vec{x}) > 0\} \tag{12.18}$$

$$\frac{\partial u^+}{\partial n} = 0 \text{ on } \{\vec{x} \mid \phi(\vec{x}) = 0\} \tag{12.19}$$

$$u^- - u_0 = \nu \Delta u^- \text{ on } \{\vec{x} \mid \phi(\vec{x}) < 0\} \tag{12.20}$$

$$\frac{\partial u^-}{\partial n} = 0 \text{ on } \{\vec{x} \mid \phi(\vec{x}) = 0\}. \tag{12.21}$$

These two sets of elliptic boundary value problems will have a smoothing and denoising effect on the image, but only inside homogeneous regions, not across edges. The Euler-Lagrange equations for ϕ, using gradient descent in artificial time, as usual, is

$$\frac{\partial \phi}{\partial t} = \delta_\epsilon(\phi)\left(\mu \nabla \cdot \left(\frac{\nabla \phi}{|\nabla \phi|}\right) - |u^+ - u_0|^2 + |u^- - u_0|^2\right. \tag{12.22}$$
$$\left. - \nu |\nabla u^+|^2 + \nu |\nabla u^-|^2\right).$$

The new numerical challenge is to obtain the numerical solution of the set of elliptic boundary value problems in equations (12.18) to (12.21) for u^- and u^+ in multiply connected regions. This is done by first extending u^- on $\{\vec{x} \mid \phi(\vec{x}) > 0\}$ while retaining boundary conditions, and similarly for u^+. There are several methods suggested; see [37] for a brief description. (One is the ghost fluid method of Fedkiw et al. [63].) See [37] for some interesting results.

The last extension is to get several, or indeed many, different regions corresponding to different level set functions. The idea is as follows. Based on the four color theorem, we can "color" all regions in a partition using only four colors such that any two adjacent regions have different colors. Therefore, using two level set functions we can identify the four colors by the four possibilities $\phi_i > 0$, $\phi_i < 0$, $i = 1, 2$. This automatically gives a segmentation of the image. However, as we shall see below, this modifies the minimization problem a bit.

As above, the link between the four regions can be made by introducing four functions u^{++}, u^{+-}, u^{-+}, and u^{++} in an obvious fashion:

$$u(\vec{x}) = \begin{cases} u^{++}(\vec{x}) & \text{if } \phi_1(\vec{x}), \phi_2(\vec{x}) > 0, \\ u^{+-}(\vec{x}) & \text{if } \phi_1(\vec{x}) > 0 > \phi_2(\vec{x}), \\ u^{-+}(\vec{x}) & \text{if } \phi_1(\vec{x}) < 0 < \phi_2(\vec{x}), \\ u^{--}(\vec{x}) & \text{if } \phi_1(\vec{x}), \phi_2(\vec{x}) < 0. \end{cases} \tag{12.23}$$

This gives us

$$u = u^{++} H(\phi_1) H(\phi_2) + u^{+-} H(\phi_1)(1 - H(\phi_2)) \tag{12.24}$$
$$+ u^{-+}(1 - H(\phi_1)) H(\phi_2) + u^{--}(1 - H(\phi_1))(1 - H(\phi_2)).$$

The energy in level set formulation based on the Mumford-Shah functional is

$$F(u, \phi) = \int |u^{++} - u_0|^2 H(\phi_1) H(\phi_2)\, d\vec{x} \tag{12.25}$$
$$+ \nu \int |\nabla u^{++}|^2 H(\phi_1) H(\phi_2)\, d\vec{x}$$
$$+ \int |u^{+-} - u_0|^2 H(\phi_1)(1 - H(\phi_2))\, d\vec{x}$$
$$+ \nu \int |\nabla u^{+-}|^2 H(\phi_1)(1 - H(\phi_2))\, d\vec{x}$$

$$+ \int |u^{-+} - u_0|^2 (1 - H(\phi_1)) H(\phi_2) \, d\vec{x}$$

$$+ \nu \int |\nabla u^{-+}|^2 (1 - H(\phi_1)) H(\phi_2) \, d\vec{x}$$

$$+ \int |u^{--} - u_0|^2 (1 - H(\phi_1))(1 - H(\phi_2)) \, d\vec{x}$$

$$+ \nu \int |\nabla u^{--}|^2 (1 - H(\phi_1))(1 - H(\phi_2)) \, d\vec{x}$$

$$+ \mu \int |\nabla H(\phi_1)| \, d\vec{x} + \mu \int |\nabla H(\phi_2)| \, d\vec{x}.$$

As the authors themselves note in [37], the last expression $\int |\nabla H(\phi_1)| \, d\vec{x} + \int |\nabla H(\phi_2)| \, d\vec{x}$ is not the length of the free boundary. However, it is certainly between one and two times that quantity. Some segments are counted once, some twice. However, this releases the Mumford-Shah functional from the well-known restriction that only 120°-angle functions are possible, within the class of multiple junctions. If one wishes to minimize the precise term proportional to length in a multiphase problem, one can use the technique involving constraints and more level set functions that was introduced by Zhao et al. [175]. For the segmentation active contours problem described here, this technique involving 2 (or $[\log_2 n]$ if one does not use the four color result and n is the number of separate regions desired) level set functions seems to work quite well for the piecewise constant case (see [36]).

The Euler-Lagrange equations for the four u functions are as in the two-phase case which means that they decouple. The time-dependent coupled gradient descent equations for ϕ_1 and ϕ_2 are easily solved with very simple changes over the two-phase, one-ϕ case.

In Figures 12.6 and 12.7 the vector-valued active contour model from [32] is used, where objects are recovered from combined channels with missing information in each channel. In Figures 12.9, 12.10, and 12.11, the piecewise-constant four-phase segmentation model from [36] is used, as a particular case of the piecewise-smooth four-phase model from [37]. A similar result from [36] is shown in Figure 12.12, where triple junctions are also detected in a color image, with the piecewise-constant segmentation model using three level set functions.

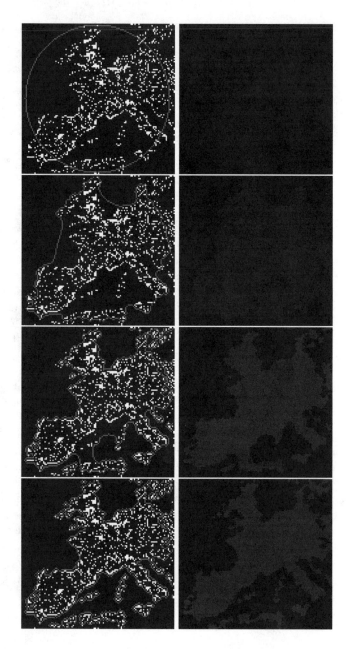

Figure 12.2. Europe night-lights [34].

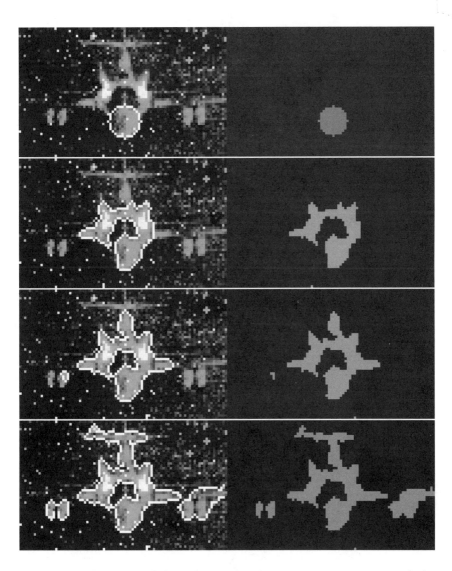

Figure 12.3. Detection of the contours of a plane in a noisy environment [34].

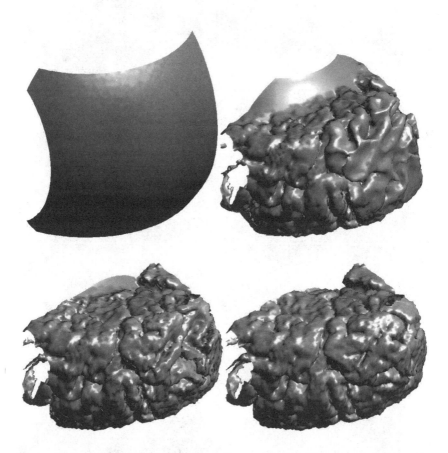

Figure 12.4. Evolution of an active surface using the 3D version of the active contour without edges from [33] on volumetric MRI brain data. We show here only a $61 \times 61 \times 61$ cube from the 3D calculations performed on a larger domain containing the brain.

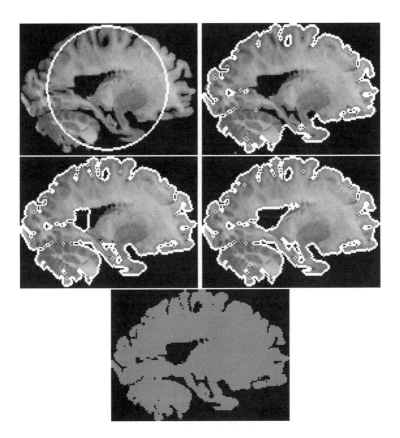

Figure 12.5. Cross-sections of the previous 3D calculations showing the evolving contour and the final segmentation on a slice of the volumetric image. We illustrate here how interior boundaries are automatically detected.

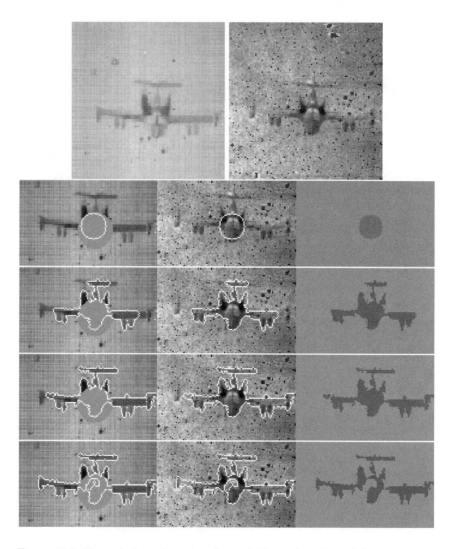

Figure 12.6. Numerical results using the multichannel version of the active contour model without edges (from [32]) to detect the full contour of an airplane from two channels. Note that channel 1 has an occlusion, while channel 2 is noisy.

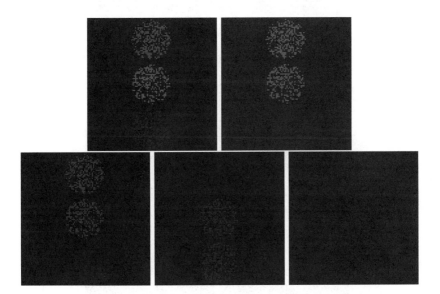

Figure 12.7. Color image, its gray-level version, and the three RGB channels. (See also color figure, Plate 3.)

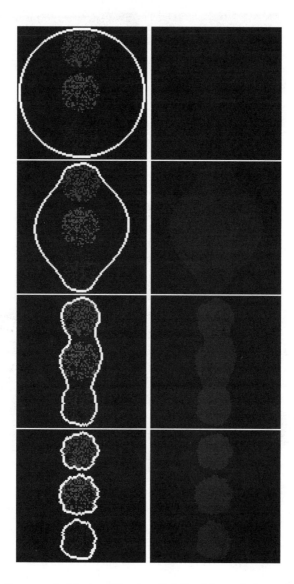

Figure 12.8. Recovered objects without well-defined boundaries, using the multi-channel version of the active contour model without edges from [32]. The three objects could not be recovered using only one channel or the intensity image. (See also color figure, Plate 4.)

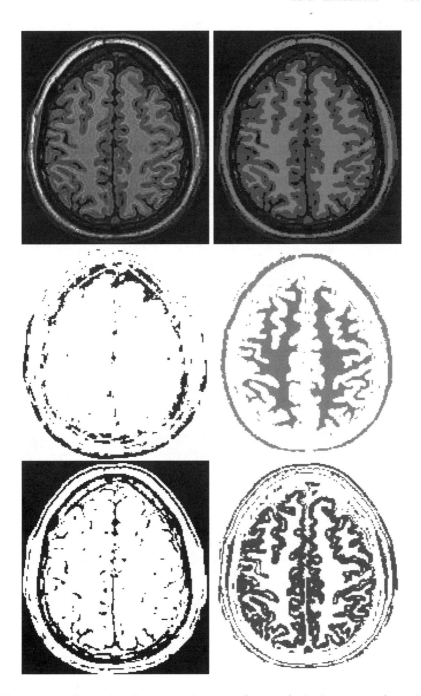

Figure 12.9. Original and segmented images (top row); final segments (second and third rows) [36].

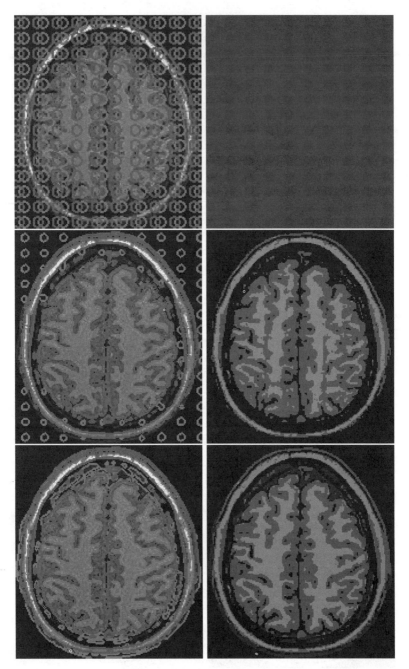

Figure 12.10. Evolution of the four-phase segmentation model using two level set functions. Left: the evolving curves. Right: corresponding piecewise-constant segmentations. Initially, we seed the image with small circles to speed up the numerical calculation [36].

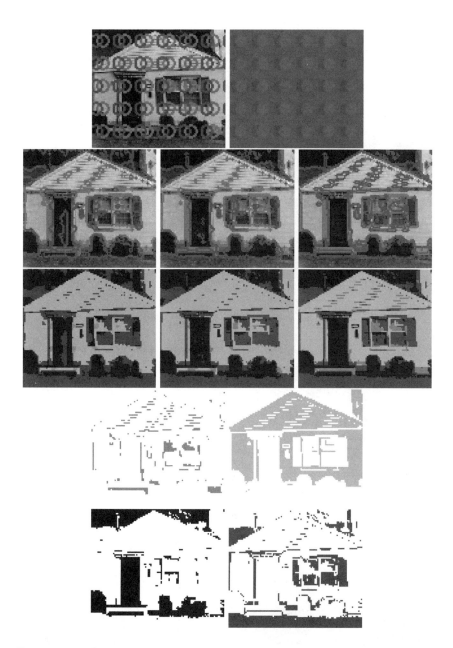

Figure 12.11. Segmentation of an outdoor picture using two level set functions and four phases. In the bottom row we show the four segments obtained [36].

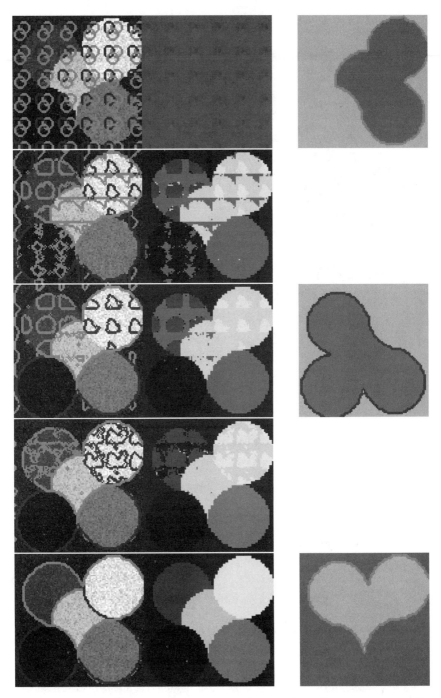

Figure 12.12. Color picture with junctions. Three level set functions representing up to eight regions. Six segments are detected. We show the final zero level sets [36]. (See also color figure, Plate 5.)

13
Reconstruction of Surfaces from Unorganized Data Points

13.1 Introduction

Surface reconstruction from an unorganized data set is very challenging. The problem is ill-posed, i.e., there is no unique solution. Furthermore, the ordering or connectivity of the data set and the topology of the real surface can be rather complicated. A desirable reconstruction procedure should be able to deal with complicated topology and geometry as well as noise and nonuniformity of the data to construct a surface that is a good approximation of the data set and has some smoothness (regularity). Moreover, the reconstructed surface should have a representation and data structure that is not only good for static rendering but also good for deformation, animation, and other dynamic operations on surfaces.

For parametric surfaces such as NURBS (see Peigl and Tiller [128] or Rogers [137]), the reconstructed surface is smooth, and the data set can be nonuniform. However, this requires one to parameterize the data set in a nice way such that the reconstructed surface is a graph in the parameter space. The parameterization and patching can be difficult for surface reconstruction from an arbitrary data set. Also, noise in the data is difficult to deal with. Another popular approach is to reconstruct a triangulated surface using Delaunay triangulations and Voronoi diagrams. The reconstructed surface is typically a subset of the faces of the Delaunay triangulations. A lot of work has been done along these lines (see, for example, Amenta and Bern [6], Boissonat and Cazals [17], and Edelsbrunner [58]), and efficient algorithms are available to compute Delaunay triangulations and Voronoi diagrams. Although this approach is more versatile in that it can deal with more general data sets, the constructed surface is only piecewise linear, and it is difficult to handle nonuniform and noisy data. Furthermore, the tracking of large deformations and topological changes can be difficult using explicit surfaces.

Recently, implicit surfaces, or volumetric representations, have attracted significant attention. The traditional approach (see Bloomenthal et al. [16]) uses a combination of smooth basis function primitives such as blobs to find a scalar function such that all data points are close to an isocontour of that scalar function. This isocontour represents the constructed implicit surface. However, computation costs are very high for large data sets, since the construction is global, which results in solving a large linear system; i.e., the basis functions are coupled together, and a single data point change can result in globally different coefficients. This makes human interaction, incremental updates, and deformation difficult. However, recently, Carr et al. [26] used polyharmonic radial basis functions (RBF) to model large data sets by a single RBF. The key new idea here is the use of the Fast Multipole Method (FMM) of Greengard and Rokhlin [76] to greatly reduce the storage and computational costs of the method. Another crucial idea is the use of off-surface points on both sides of the point cloud. However, the ability to interpolate curves and surface patches, the ability to do dynamic deformation, the performance on coarse data sets, and the speed of the method all seem to be inferior to the method we describe below. On the other hand, the method proposed by [26] does give an analytic, grid-free expression and exact control of the filtering error.

Zhao et al. [177, 176] proposed a new weighted minimal surface model based on variational formulations. Only the unsigned distance function to the data set is used, and the reconstructed surface is smoother than piecewise linear. The formulation is a regularization that is adaptive to the local sampling density that can keep sharp features if a local sampling condition is satisfied. The method handles noisy as well as nonuniform data and works well in three spatial dimensions.

13.2   The Basic Model

Let S denote a general data set, which can include data points, curves, and pieces of surfaces. Define $d(\vec{x}) = \text{dist}(\vec{x}, S)$ to be the distance function to S. The following surface energy is defined for the variational formulation:

$$E(\Gamma) = \left[\int_{\Gamma} d^p(\vec{x}) \, ds \right]^{\frac{1}{p}}, \quad 1 \leq p \leq \infty, \tag{13.1}$$

where Γ is an arbitrary surface and ds is the surface area. The energy functional is independent of parameterization and is invariant under rotation and translation. When $p = \infty$, $E(\Gamma)$ is the value of the distance of the point $\vec{x}$ on Γ furthest from S. We take the local minimizer of our energy functional, which mimics a weighted minimal surface or an elastic membrane attached to the data set, to be the reconstructed surface.

The gradient flow of the energy functional in equation (13.1) is

$$\frac{d\Gamma}{dt} = -\left[\int_{\Gamma} d^p(\vec{x}) \, ds \right]^{\frac{1}{p}-1} d^{p-1}(\vec{x}) \left[\nabla d(\vec{x}) \cdot \vec{N} + \frac{1}{p} d(\vec{x})\kappa \right] \vec{N}, \tag{13.2}$$

and the minimizer or steady-state solution of the gradient flow satisfies the Euler-Lagrange equation

$$d^{p-1}(\vec{x}) \left[\nabla d(\vec{x}) \cdot \vec{N} + \frac{1}{p} d(\vec{x})\kappa \right] = 0. \tag{13.3}$$

We see a balance between the attraction $\nabla d(\vec{x}) \cdot \vec{N}$ and the surface tension $d(\vec{x})\kappa$. Moreover, the nonlinear regularization due to surface tension has a desirable scaling $d(\vec{x})$. Thus the reconstructed surface is more flexible in the region where the sampling density is high and is more rigid in the region where the sampling density is low. We start with an initial surface that encloses all the data and follow the gradient flow in equation (13.2). When $p = 1$, the surface energy defined in equation (13.1) has the dimension of volume and the gradient flow in equation (13.2) is scale-invariant. In practice we find that $p = 1$ is a good choice.

We use the same motion law for all level sets of the level set function, which results in a morphological partial differential equation. The level set formulation becomes

$$\frac{\partial \phi}{\partial t} = \left[\nabla d \cdot \frac{\nabla \phi}{|\nabla \phi|} + d\nabla \cdot \frac{\nabla \phi}{|\nabla \phi|} \right] |\nabla \phi|. \tag{13.4}$$

If the data contain noise, we can use a simple postprocessing for the implicit surface. There are many ways to view this process, derived by Whitaker [172], but perhaps the most relevant here is based on TV denoising of images described in Chapter 11. Consider ϕ_0, the level set function whose zero isocontour is the surface we wish to smooth. Then we let $u_0 = H(\phi_0)$ (H is the Heaviside function) be the noisy image, which we input into the TV denoising algorithm. Then we minimize

$$\mu TV(H(\phi)) + \frac{1}{2} \int (H(\phi) - H(\phi_0))^2 d\vec{x},$$

where $\mu > 0$ is the regularization parameter that balances between fidelity and regularization. The variational level set method of Zhao et al. [175] gives

$$\phi_t = [\mu\kappa - (H(\phi) - H(\phi_0))]|\nabla \phi|, \tag{13.5}$$

and we take ϕ_0 as the initial guess.

13.3   The Convection Model

The evolution equation (13.2) involves the mean curvature of the surface, and it is a nonlinear parabolic equation. A time-implicit scheme is not currently available. A stable time-explicit scheme requires a restrictive-time step size, $\Delta t = O(\triangle x^2)$. Thus it is desirable to have an efficient algorithm to find a good approximation before we start the gradient flow for the minimal surface. We propose the following physically motivated convection model for this purpose.

If a velocity field is created by a potential field $\mathcal{F}$, then $\vec{v} = -\nabla \mathcal{F}$. In our convection model the potential field is the distance function $d(\vec{x})$ to the data set S. This leads to the convection equation

$$\frac{\partial \phi}{\partial t} = \nabla d(\vec{x}) \cdot \nabla \phi. \tag{13.6}$$

For a general data set S, a particle will be attracted to its closest point in S unless the particle is located an equal distance from two or more data points. The set of equal distance points has measure zero. Similarly, points on our surface, except those equal distance points, are attracted by their closest points in the data set. The ambiguity at those equal distance points is resolved by adding a small surface tension force, which automatically exists as numerical viscosity in our finite difference schemes. Those equal distance points on the curve or surface are dragged by their neighbors, and the whole curve or surface is attracted to the data set until it reaches a local equilibrium, which is a polygon or polyhedron whose vertices belong to the data set as the viscosity tends to zero.

The convection equation can be solved using a time step $\Delta t = O(\triangle x)$, leading to significant computational savings over typical parabolic $\Delta t = O(\triangle x^2)$ time-step restrictions. The convection model by itself very often results in a good surface construction.

13.4   Numerical Implementation

There are three key numerical ingredients in our implicit surface reconstruction. First, we need a fast algorithm to compute the distance function to an arbitrary data set on rectangular grids. Second, we need to find a good initial surface for our gradient flow. Third, we have to solve time-dependent partial differential equations for the level set function.

We can use an arbitrary initial surface that contains the data set such as a rectangular bounding box, since we do not have to assume any a priori knowledge for the topology of the reconstructed surface. However, a good initial surface is important for the efficiency of our method. We start from any initial exterior region that is a subset of the true exterior region. All grid points that are not in the initial exterior region are labeled as interior points. Those interior grid points that have at least one exterior neighbor are labeled as temporary boundary points. Then we use the following procedure to march the temporary boundary inward toward the data set. We put all the temporary boundary points in a heap-sort binary tree structure, sorting according to distance values. Take the temporary boundary point that has the largest distance (on the heap top) and check to see whether it has an interior neighbor that has a larger or equal distance value. If it does not have such an interior neighbor, turn this temporary boundary point into an exterior point, take this point out of the heap, add all this point's interior neighbors into the heap, and re-sort according to distance values. If it does have such an interior neighbor, we turn this temporary boundary point into a final boundary point, take it out of the heap, and re-sort the heap. None of its neighbors are added to the heap. We repeat this procedure on the temporary boundary points until the the maximum distance of the temporary boundary points is smaller than some tolerance, e.g., the size of a grid cell, which means that all the temporary boundary points in the heap are close enough to the data set. Finally, we turn these temporary boundary points into the final set of boundary points, and our tagging procedure is finished. Since we visit each interior grid point at most once, the procedure will be completed in no more than $O(N \log N)$ operations, where $\log N$ comes from the heap-sort algorithm. Moreover, since the maximum distance for the boundary heap is strictly decreasing, we can prove that those interior points that have a distance no smaller than the maximum distance of the temporary boundary heap at any time will remain as interior points; i.e., there is a nonempty interior region when the tagging algorithm is finished. We can also show that at least one of the final boundary points is within the tolerance distance to the data set.

Starting from an arbitrary exterior region that is a subset of the final exterior region, the furthest point on the temporary boundary is tangent to a distance contour and does not have an interior point that is farther away. The furthest point will be tagged as an exterior point, and the boundary will move inward at that point. Now another point on the temporary boundary becomes the furthest point, and hence the whole temporary boundary moves inward. After a while the temporary boundary is close to a distance contour and moves closer and closer to the data set, following the distance contours until the distance contours begin to break into spheres around data points. The temporary boundary point at the breaking point of the distance contour, which is equally distant from distinct data points, will have neighboring interior points that have a larger distance. So this temporary boundary point will be tagged as a final boundary point by our procedure, and the temporary boundary will stop moving inward at this breaking point. The temporary boundary starts deviating from the distance contours and continues moving closer to the data set until all temporary boundary points either have been tagged as final boundary points or are close to the data points. The final boundary is approximately a polyhedron with vertices belonging to the data set.

Figure 13.1 shows the reconstruction of a torus with missing data. The hole is filled nicely with a patch of minimal surface. Figure 13.2 shows the reconstruction of a rat brain from MRI data, which is both noisy and highly nonuniform between slices. Next we show the reconstruction of a dragon on a $300 \times 212 \times 136$ grid using high-resolution data in Figure 13.3(a) and much lower resolution data in figure 13.3(b). Figure 13.4 shows the reconstruction of a statuette of the Buddha on two different grids using the same data set composed of 543,652 points.

Other extensions are possible. Suppose we are given values of the normal to the surface at the same or different set S' of points. The first step, analogous to the fast computation of unsigned distance, is to construct a unit vector defined throughout the grid that interpolates this set. One possibility involves the construction of a harmonic map, which is easier than in sounds using any of the techniques developed by Vese and Osher [170], Alouges [4], E and Wang [57], or Tang et al. [162]. Given this unit vector $\vec{N}(\vec{x})$ we add to our energy $E(\Gamma)$ another quantity $cE'(\Gamma)$, where $c > 0$ is a constant whose dimension is length,

$$E'(\Gamma) = \left(\int_{\Gamma} \left(1 - \vec{N} \cdot \left(\frac{\nabla \phi}{|\nabla \phi|} \right) \right)^p ds \right)^{\frac{1}{p}}. \tag{13.7}$$

Again using our variational level set calculus, we see that the gradient descent evolution associated with equation (13.7) is

$$\phi_t = |\nabla \phi| \left(\nabla \cdot \frac{\nabla \phi}{|\nabla \phi|} - \nabla \cdot \vec{n} \right) \tag{13.8}$$

for $p = 1$. See Burchard et al. [21].

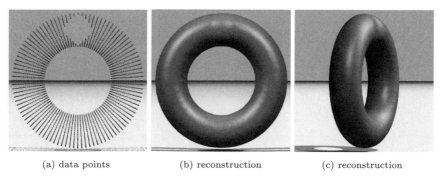

(a) data points   (b) reconstruction   (c) reconstruction

Figure 13.1. Hole-filling for a torus.

(a) data points   (b) initial guess   (c) final reconstruction

Figure 13.2. Reconstruction of a rat brain from data of MRI slices.

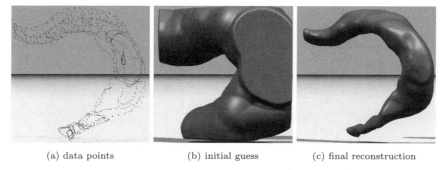

(a) 437,645 points   (b) 100,250 points

Figure 13.3. Reconstruction of a dragon using data sets of different resolution on a $300 \times 212 \times 136$ grid.

(a) 146x350x146 grid   (b) 63x150x64 grid

Figure 13.4. Reconstruction of a "Happy Buddha" from 543,652 data points on different grid resolutions.

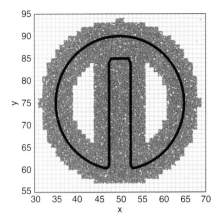

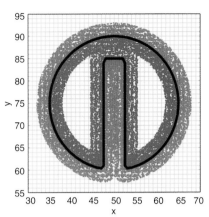

Plate 1. (Figure 9.5). Initial placement of both types of particles on both sides of the interface.

Plate 2. (Figure 9.5). Particle positions after the initial attraction step is used to place them on the appropriate side of the interface.

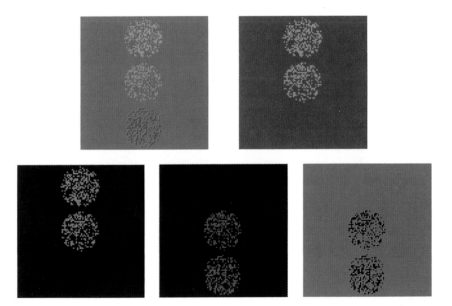

Plate 3 (Figure 12.7). Color image, its gray-level version, and the three RGB channels.

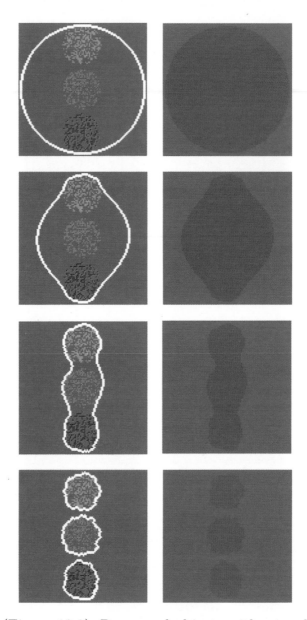

Plate 4 (Figure 12.8). Recovered objects without well-defined boundaries, using the multi-channel version of the active contour model without edges from [32]. The three objects could not be recovered using only one channel or the intensity image.

Plate 5 (Figure 12.12). Color picture with junctions. Three level set functions representing up to eight regions. Six segments are detected. We show the final zero level sets [36].

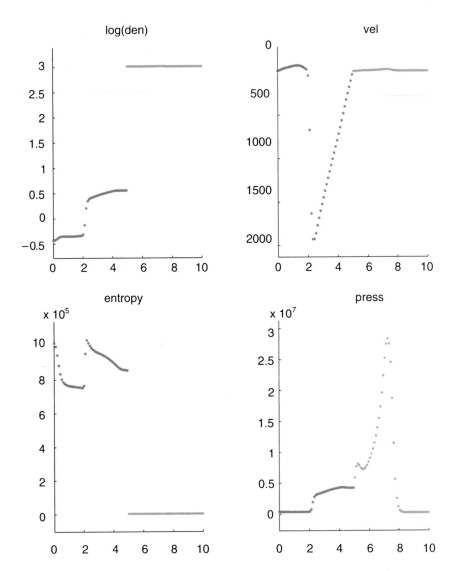

Plate 6 (Figure 15.9). The gamma-law gas is depicted in red, while the stiff Tait equation of state water is depicted in green. Note that the log of the density is shown, since the density ratio is approximately 1000 to 1. This calculation uses only 100 grid cells.

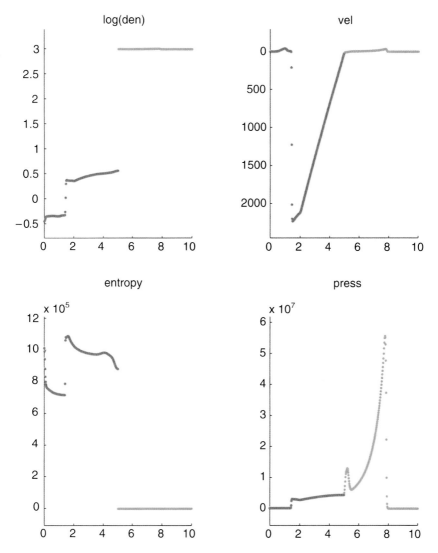

Plate 7 (Figure 15.10). This is the same calculation as in Figure 15.9, except that 500 grid cells are used.

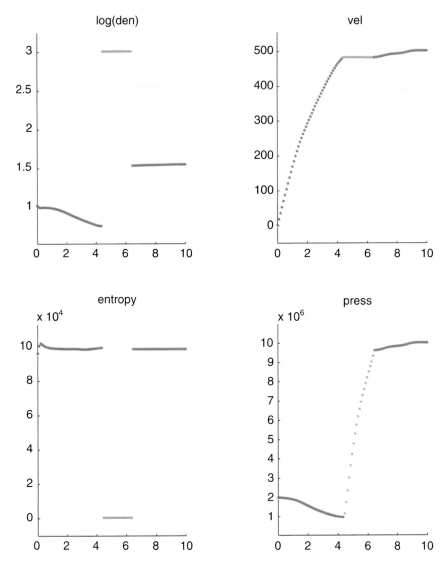

Plate 8 (Figure 15.11). In this calculation two interfaces are present, since the air surrounds the water on both sides. This calculation uses only 100 grid cells.

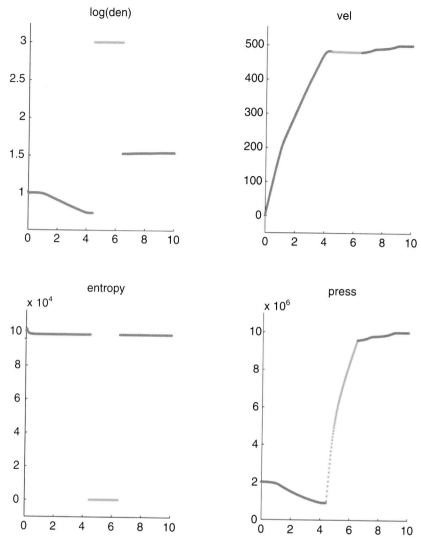

Plate 9 (Figure 15.12). This is the same calculation as in Figure 15.11, except that 500 grid cells are used.

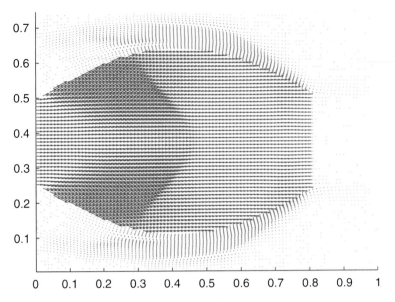

Plate 10 (Figure 17.1). A shock wave propagating through a gas bounded on top and bottom by Lagrangian materials with strength.

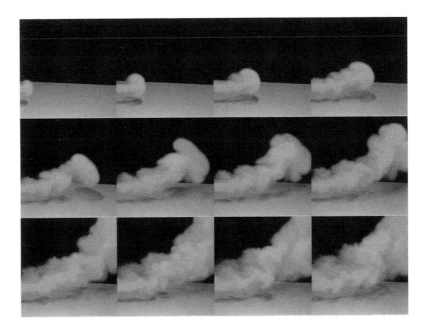

Plate 11 (Figure 18.1). A warm smoke plume injected from left to right rises under the influence of buoyancy.

Plate 12. (Figure 18.2). Small-scale eddies are generated as smoke flows past a sphere.

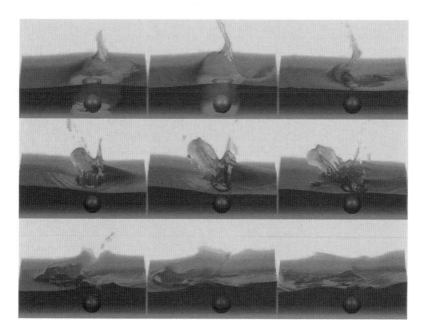

Plate 13. (Figure 19.1). A splash is generated as a sphere is thrown into the water.

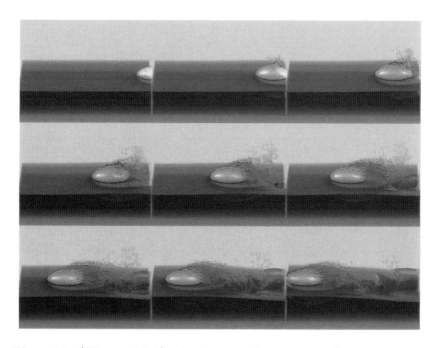

Plate 14. (Figure 19.2). An interesting spray effect is generated as a slightly submerged ellipse slips through the water.

Plate 15. (Figure 19.3). A thin water sheet is generated by a sphere thrown into the water.

Plate 16 (Figure 19.4). Pouring water into a cylindrical glass using the particle level set method.

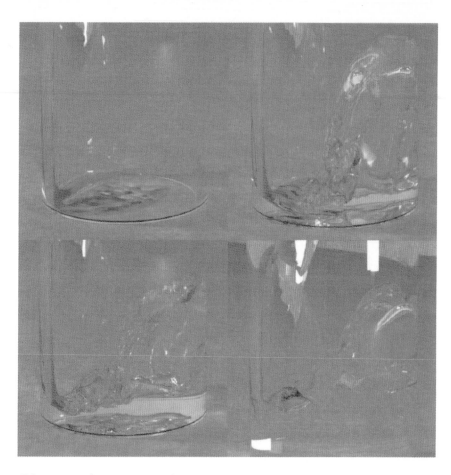

Plate 17. (Figure 19.5). Pouring water into a cylindrical glass using the particle level set method.

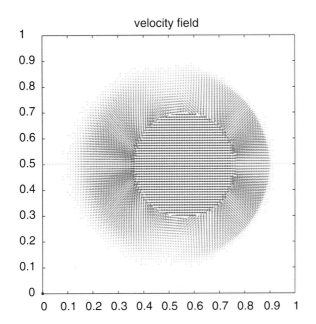

Plate 18. (Figure 20.1). An incompressible droplet traveling to the right in a compressible gas flow. Note the lead shock wave.

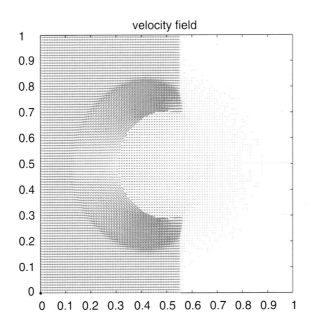

Plate 19. (Figure 20.2). A shock wave impinging on an incompressible droplet producing a reflected wave and a (very) weak transmitted wave.

Plate 20. (Figure 21.2). A water drop falls through the air into the water. Surface tension forces cause the spherically shaped region at the top of the water jet in the last frame.

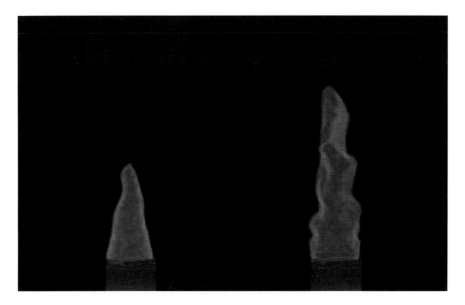

Plate 21. (Figure 22.4). Typical blue cores rendered using the zero isocontour of the level set function.

Plate 22. (Figure 22.5). The density ratio of the unburnt to burnt gas is increased from left to right, illustrating the effect of increased expansion.

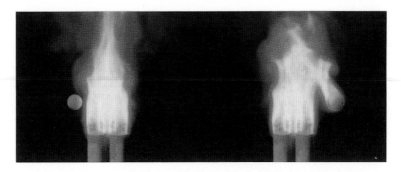

Plate 23. (Figure 22.6). A flammable ball catches on fire as it passes through a flame.

Plate 24. (Figure 22.7). Campfire with realistic lighting of the surrounding rocks.

Part IV
Computational Physics

While the field of computational fluid dynamics is quite broad, a large portion of it is dedicated to computations of compressible flow, incompressible flow, and heat flow. In fact, these three classes of problems can be thought of as the basic model problems for hyperbolic, elliptic, and parabolic partial differential equations. Volumes have been filled with both broad and detailed work dedicated to these important flow fields; see, for example, [86], [87], [7], [8], and the references therein. One might assume that there is little left to add to the understanding of these problems. In fact, many papers are now concerned with smaller details, e.g., the number of grid points in a shock wave or contact discontinuity. Other papers are devoted to rarely occurring pathologies, e.g., slow-moving shock waves and shock overheating at solid wall boundaries.

One area where significant new ideas are still needed is multicomponent flow, for example, multicomponent compressible flow where different fluids have different equations of state, multicomponent incompressible flow where different fluids have different densities and viscosities with surface tension forces at the interface, and Stefan-type problems where the individual materials have different thermal conductivities. These problems have interfaces separating the different materials, and special numerical techniques are required to treat the interface. The most commonly used are front tracking, volume of fluid, and level set methods. In the next three chapters we discuss the use of level set methods for the canonical equations of computational fluid dynamics.

The first chapter discusses basic one-phase compressible flow, and the subsequent chapter shows how the level set method can be used for two-phase compressible flow problems where the equations of state differ across the interface. Then, the use of level set techniques for deflagration and detonation discontinuities is discussed in the third chapter. The fourth chapter introduces techniques for coupling an Eulerian grid to a Lagrangian grid. This is useful, for example, in compressible solid/fluid structure problems. After this, we turn our attention to incompressible flow with a chapter focused on the basic one-phase equations including a computer graphics simulation of smoke, a chapter on level set techniques for free surface flows including a computer graphics simulation of water, and finally a chapter on fully two-phase incompressible flow. We wrap up this incompressible flow material with a chapter on incompressible flames, including applications in computer graphics, and a chapter on techniques for coupling a compressible fluid to an incompressible fluid. Finally, we turn our attention to heat flow with a chapter discussing the heat equation and a chapter discussing level set techniques for solving Stefan problems.

14

Hyperbolic Conservation Laws and Compressible Flow

We begin this chapter by addressing general systems of hyperbolic conservation laws including numerical techniques for computing accurate solutions to them. Then we discuss the equations for one-phase compressible flow as an example of a system of hyperbolic conservation laws.

14.1 Hyperbolic Conservation Laws

A continuum physical system is described by the laws of conservation of mass, momentum, and energy. That is, for each conserved quantity, the rate of change of the total amount in some region is given by its flux (convective or diffusive) through the region boundary, plus whatever internal sources exist. The integral form of this conservation law is

$$\frac{d}{dt} \int_\Omega u \, dV + \int_{\partial\Omega} \vec{f}(u) \cdot dA = \int_\Omega s(u) \, dV \tag{14.1}$$

where u is the density of the conserved quantity, $\vec{f}(u)$ is the flux, and $s(u)$ is the source rate. By taking Ω to be an infinitesimal volume and applying the divergence theorem, we get the differential form of the conservation law,

$$\frac{\partial u}{\partial t} + \nabla \cdot \vec{f}(u) = s(u), \tag{14.2}$$

which is the basis for the numerical modeling of continuum systems. A physical system can be described by a system of such equations, i.e., a system of conservation laws. These also form the basis for their numerical modeling.

A conserved quantity, such as mass, can be transported by convective or diffusive fluxes. The distinction is that diffusive fluxes are driven by gradients, while convective fluxes persist even in the absence of gradients. For most flows where compressibility is important, e.g., flows with shock waves, one needs to model only the convective transport and can ignore diffusion (mass diffusion, viscosity and thermal conductivity) as well as the source terms (such as chemical reactions, atomic excitations, and ionization processes). Moreover, convective transport requires specialized numerical treatment, while diffusive and reactive effects can be treated with standard numerical methods, such as simple central differencing, that are independent of those for the convective terms. Stiff reactions, however, can present numerical difficulties; see, for example, Colella et al. [50]. Conservation laws with only convective fluxes are known as hyperbolic conservation laws. A vast array of physical phenomena are modeled by such systems, e.g., explosives and high-speed aircraft.

The important physical phenomena exhibited by hyperbolic conservation laws are bulk convection, waves, contact discontinuities, shocks, and rarefactions. We briefly describe the physical features and mathematical model equations for each effect, and most importantly note the implications they have on the design of numerical methods. For more details on numerical methods for conservation laws, see, e.g., LeVeque [105] and Toro [164].

14.1.1   Bulk Convection and Waves

Bulk convection is simply the bulk movement of matter, carrying it from one spot to another, like water streaming from a hose. Waves are small-amplitude smooth disturbances that are transmitted through the system without any bulk transport like ripples on a water surface or sound waves through air. Whereas convective transport occurs at the gross velocity of the material, waves propagate at the "speed of sound" in the system (relative to the bulk convective motion of the system). Waves interact by superposition, so that they can either cancel out (interfere) or enhance each other.

The simplest model equation that describes bulk convective transport is the linear convection equation

$$u_t + \vec{v} \cdot \nabla u = 0, \qquad (14.3)$$

where $\vec{v}$ is a constant equal to the convection velocity. The solution to this is simply that u translates at the constant speed $\vec{v}$. This same equation can also be taken as a simple model of wave motion if u is a sine wave and $\vec{v}$ is interpreted as the speed of sound. The linear convection equation is also an important model for understanding smooth transport in any conservation law. As long as $\vec{f}$ is smooth and u has no jumps in it, the general scalar conservation law

$$u_t + \nabla \cdot \vec{f}(u) = 0 \tag{14.4}$$

can be rewritten as

$$u_t + \vec{f'}(u) \cdot \nabla u = 0, \tag{14.5}$$

where $\vec{f'}(u)$ acts as a convective velocity. That is, locally in smooth parts of the flow, a conservation law behaves like bulk convection with velocity $\vec{f'}(u)$. This is called the local characteristic velocity of the flow.

Bulk convection and waves are important because they imply that signals propagate in definite directions at definite speeds. This is in contrast to diffusion, which propagates signals in all directions at arbitrarily large speeds depending on the severity of the driving gradients. Thus we anticipate that suitable numerical methods for hyperbolic systems will also have directional biases in space, which leads to the idea of upwind differencing and a definite relation between the space and time steps (discrete propagation speed), which will roughly be that the discrete propagation speed $\Delta x / \Delta t$ must be at least as large as the physical propagation speeds (characteristic speeds) in the problem. The general form of this relation is called the Courant-Friedrichs-Lewy (CFL) restriction.

Wave motion and bulk convection do not create any new sharp features in the flow. The other remaining phenomena are all special because they involve discontinuous jumps in the transported quantities. Because smooth features can be accurately represented by a polynomial interpolation, we expect to be able to develop numerical methods of extremely high accuracy for the wave and convective effects. Conversely, since jump functions are poorly represented by polynomials, we expect little accuracy and perhaps great difficulty in numerically approximating the discontinuous phenomena.

14.1.2 Contact Discontinuities

A contact discontinuity is a persistent discontinuous jump in mass density moving by bulk convection through a system. Since there is negligible mass diffusion, such a jump persists. These jumps usually appear at the point of contact of different materials; for example, a contact discontinuity separates oil from water. Contacts move at the local bulk convection speed, or more generally the characteristic speed, and can be modeled by using step-function initial data in the bulk convection equation (14.3). Since contacts are simply a bulk convection effect, they retain any perturbations they receive. Thus we expect contacts to be especially sensitive to numerical methods; i.e., any spurious alteration of the contact will tend to persist and accumulate.

14.1.3 Shock Waves

A shock is a spatial jump in material properties, like pressure and temperature, that develops spontaneously from smooth distributions and then persists. The shock jump is self-forming and also self-maintaining. This is unlike a contact discontinuity, which must be put in the system initially and will not resharpen itself if it is smeared out by some other process. Shocks develop through a feedback mechanism in which strong impulses move faster than weak ones, and thus tend to steepen themselves up into a "step" profile as they travel through the system. Familiar examples are the "sonic boom" of a jet aircraft and the "bang" from a gun. These sounds are our perceptions of a sudden jump in air pressure.

The simplest model equation that describes shock formation is the one-dimensional Burgers' equation

$$u_t + \left(\frac{u^2}{2}\right)_x = 0, \tag{14.6}$$

which looks like the convection equation (14.3) with a nonconstant convective speed of u, i.e., $u_t + u u_x = 0$. Thus larger u values move faster, and they will overtake smaller values. This ultimately results in the development of, for example, a right-going shock if the initial data for u constitute any positive, decreasing function.

Shocks move at a speed that is not simply related to the bulk flow speed or characteristic speed, and they are not immediately evident from examining the flux, in contrast to contacts. Shock speed is controlled by the difference between influx and outflux of conserved quantity into the region. Specifically, suppose a conserved quantity u with conservation law

$$u_t + f(u)_x = 0 \tag{14.7}$$

has a step function profile with constant values extending both to the left, u_L, and to the right, u_R, with a single shock jump transition in between moving with speed s. Then the integral form of the conservation law (14.1), applied to any interval containing the shock, gives the relation

$$s(u_R - u_L) = f(u_R) - f(u_L), \tag{14.8}$$

which is just another statement that the rate at which u appears, $s(u_R - u_L)$, in the interval of interest is given by the difference in fluxes across the interval. Thus we see that the proper speed of the shock is directly determined by conservation of u via the flux f. This has an important implication for numerical method design; namely, a numerical method will "capture" the correct shock speeds only if it has "conservation form," i.e., if the rate of change of u at some node is the difference of fluxes that are accurate approximations of the real flux f.

The self-sharpening feature of shocks has two implications for numerical methods. First, it means that even if the initial data are smooth, steep gradients and jumps will form spontaneously. Thus, our numerical method must be prepared to deal with shocks even if none are present in the initial data. Second, there is a beneficial effect from self-sharpening, because modest numerical errors introduced near a shock (smearing or small oscillations) will tend to be eliminated, and will not accumulate. The shock is naturally driven toward its proper shape. Because of this, computing strong shocks is mostly a matter of having a conservative scheme in order to get their speed correct.

14.1.4 Rarefaction Waves

A rarefaction is a discontinuous jump or steep gradient in properties that dissipates as a smooth expansion. A common example is the jump in air pressure from outside to inside a balloon, which dissipates as soon as the balloon is burst and the high-pressure gas inside is allowed to expand. Such an expansion also occurs when the piston in an engine is rapidly pulled outward from the cylinder. The expansion (density drop) associated with a rarefaction propagates outward at the sound speed of the system, relative to the underlying bulk convection speed. A rarefaction can be modeled by Burgers' equation (14.6) with initial data that start out as a steep increasing step. This step will broaden and smooth out during the evolution.

A rarefaction tends to smooth out local features, which is generally beneficial for numerical modeling. It tends to diminish numerical errors over time and make the solution easier to represent by polynomials, which form the basis for our numerical representation. However, a rarefaction often connects to a smooth (e.g., constant) solution region and this results in a "corner," which is notoriously difficult to capture accurately. The main numerical problem posed by rarefactions is that of initiating the expansion. If the initial data is are perfect, symmetrical step, such as $u(x) = \text{sign}(x)$, it may be "stuck" in this form, since the steady-state Burgers' equation is satisfied identically (i.e., the flux $u^2/2$ is constant everywhere, and similarly in any numerical discretization). However, local analysis can identify this stuck expansion, because the characteristic speed u on either side points away from the jump, suggesting its potential to expand. In order to get the initial data unstuck, a small amount of smoothing must be applied to introduce some intermediate-state values that have a nonconstant flux to drive expansion. In numerical methods this smoothing applied at a jump where the effective local velocity indicates expansion should occur is called an "entropy fix," since it allows the system to evolve from the artificial low entropy initial state to the proper increased entropy state of a free expansion.

14.2   Discrete Conservation Form

To ensure that shocks and other steep gradients are captured by the scheme, i.e., that they move at the right speed even if they are unresolved, we must write the equation in a discrete conservation form. That is, a form in which the rate of change of conserved quantities is equal to a difference of fluxes. This form guarantees that we discretely conserve the total amount of the states u (e.g., mass, momentum, and energy) present, in analogy with the integral form given by equation (14.1). More important, this can be shown to imply that steep gradients or jumps in the discrete profiles propagate at the physically correct speeds; see, for example, LeVeque [105].

Usually, conservation form is derived for control volume methods, that is methods that evolve cell average values in time rather than nodal values. In this approach, a grid node x_i is assumed to be the center of a grid cell $(x_{i-1/2}, x_{i+1/2})$, and we integrate the conservation law (14.7) across this control volume to obtain

$$\bar{u}_t + f(u_{i+1/2}) - f(u_{i-1/2}) = 0, \qquad (14.9)$$

where $\bar{u}$ is the integral of u over the cell, and $u_{i\pm1/2}$ are the (unknown) values of u at the cell walls. This has the desired conservation form in that the rate of change of the cell average is a difference of fluxes. The difficulty with this formulation is that it requires transforming between cell averages of u (which are directly evolved in time by the scheme) and cell wall values of u (which must be reconstructed) to evaluate the needed fluxes. While this is manageable in one spatial dimension, in higher-dimensional problems the series of transformations necessary to convert the cell averages to cell wall quantities becomes increasingly complicated. The distinction between cell average and midpoint values can be ignored for schemes whose accuracy is no higher than second order (e.g., TVD schemes), since the cell average and the midpoint value differ by only $O(\Delta x^2)$.

Shu and Osher [150, 151] proposed a fully conservative finite difference scheme on uniform grids that directly evolves nodal values (as opposed to the cell average values) forward in time. They defined a numerical flux function $\mathcal{F}$ by the property that the real flux divergence is a finite difference of numerical fluxes

$$f(u)_x = \frac{\mathcal{F}(x + \Delta x/2) - \mathcal{F}(x - \Delta x/2)}{\Delta x} \qquad (14.10)$$

at every point x. We call $\mathcal{F}$ the numerical flux, since we require it in our numerical scheme, and also to distinguish it from the closely related "physical flux" $f(u)$. It is not obvious that the numerical flux function exists, but from relationship (14.10) one can solve for its Taylor expansion to obtain

$$\mathcal{F} = f(u) - \frac{(\Delta x)^2}{24} f(u)_{xx} + \frac{7(\Delta x)^4}{5760} f(u)_{xxxx} - \cdots, \qquad (14.11)$$

which shows that the physical and numerical flux functions are the same to second-order accuracy in Δx. Thus, a finite difference discretization can be based on

$$u_t + \frac{\mathcal{F}(x + \Delta x/2) - \mathcal{F}(x - \Delta x/2)}{\Delta x} = 0 \tag{14.12}$$

to evolve point values of u forward in time using numerical flux functions $\mathcal{F}$ at the cell walls.

14.3 ENO for Conservation Laws

14.3.1 Motivation

Essentially nonoscillatory (ENO) methods were developed to address the special difficulties that arise in the numerical solution of systems of nonlinear conservation laws. Numerical methods for these problems must be able to handle steep gradients, e.g., shocks and contact discontinuities, that may develop spontaneously and then persist in these flows. Classical numerical schemes had a tendency either to produce large spurious oscillations near steep gradients or to greatly smear out both these gradients and the fine details of the flow. The primary goal of the ENO effort was to develop a general-purpose numerical method for systems of conservation laws that has high accuracy (e.g., third order) in smooth regions and captures the motion of unresolved steep gradients without creating spurious oscillations and without the use of problem-dependent fixes or tunable parameters. The philosophy underlying the ENO methods is simple: When reconstructing a profile for use in a convective flux term, one should not use high-order polynomial interpolation across a steep gradient in the data. Such an interpolant would be highly oscillatory and ultimately corrupt the computed solution. ENO methods use an adaptive polynomial interpolation constructed to avoid steep gradients in the data. The polynomial is also biased to use data from the direction of information propagation (upwind) for physical consistency and stability.

The original ENO schemes developed by Harten et al. [81] were based on the conservative control volume discretization of the equations, which yields discrete evolution equations for grid cell averages of the conserved quantities, e.g. mass, momentum, and energy. This formulation has the disadvantage of requiring complicated transfers between cell averages and cell center nodal values in the algorithm. In particular, the transfer process becomes progressively more complicated in one, two, and three spatial dimensions. The formulation also results in space and time discretizations that are coupled in a way that becomes complicated for higher-order accurate versions. To eliminate these complications, Shu and Osher [150, 151] developed a conservative finite difference form of the ENO method that uses only nodal values of the conserved variables. Their method is faster and easier to implement than the cell-averaged formulation. In addition, the finite difference ENO method extends to higher dimensions in a dimension-by-dimension fashion, so that the one-dimensional method applies unchanged to higher-dimensional problems. We emphasize that this is *not* dimensional splitting in time, which has accuracy limitations unlike the dimension-by-dimension approach. Shu and Osher also use the method of lines for time integration, decoupling the time and space discretizations.

We consider the treatment of a one-dimensional contact discontinuity to illustrate how the method works. Assuming that the time evolution takes place exactly, each time step Δt should rigidly translate the spatial profile by the amount $v\Delta t$ as governed by equation (14.3). Spatially, the contact is initially represented by a discrete step function, i.e., nodal values that are constant at one value u_L on nodes $x_1, \ldots, x_J$, and constant at a different value u_R on all remaining nodes $x_{J+1}, \ldots, x_N$. To update the value u_i in time at a given node x_i, we first reconstruct the graph of a function $u(x)$ near x_i by interpolating nearby nodal u values, shift that $u(x)$ graph spatially by $v\Delta t$ (the exact time evolution), and then reevaluate it at the node x_i to obtain the updated u_i. We require our local interpolant be smooth at the point x_i, since in actual practice we are going to use it to evaluate the derivative term u_x there. The simplest symmetric approach to smooth interpolation near a node x_i is to run a parabola through the nodal data at x_{i-1}, x_i, and x_{i+1}. This interpolation is an accurate reconstruction of $u(x)$ in smooth regions, where it works well. However, near the jump between x_J and x_{J+1} the parabola will significantly overshoot the nodal u data by an amount comparable to the jump $u_L - u_R$, and this overshoot will show up in the nodal values once the shift is performed. Successive time steps will further enhance these spurious oscillations. This approach corresponds to standard central differencing.

To avoid the oscillations from parabolic interpolation, we could instead use a smooth linear interpolation near x_i, noting that there are two linear interpolants to choose from, namely the line through the data at nodes x_i and x_{i-1}, and the line through the data at x_i and x_{i+1}. The direction of information propagation determines which should be used. If the convection speed v is positive, the data are moving from left to right, and we use x_i and x_{i-1}. This linear interpolation based on upwind nodes will not introduce any new extrema in u as long as the shift $v\Delta t$ is less than the width of the interval $\Delta x = x_i - x_{i-1}$, which is exactly the CFL restriction on the time step. The main problem with the linear upwind biased interpolant is that it has low accuracy smearing out the jump over more and more nodes. If we naively go to higher accuracy by using a higher-order upwind biased interpolant, such as running a parabola through x_i, x_{i-1}, and x_{i-2} to advance u_i, we run into the spurious oscillation problem again. In particular, at nodes x_{J+1} and x_{J+2}, this upwind parabola will interpolate across the jump and thus have large overshoots. By forcing the parabola to cross a jump, it no longer reflects the data on the interval that will be arriving at x_{J+1} (or x_{J+2}) during the next time step.

The motivation for ENO is that we must use a higher-degree polynomial interpolant to achieve more accuracy, and it must involve the immediate upwind node to properly represent the propagation of data. But we must also avoid polluting this upwind data with spurious oscillations that come from interpolating across jumps. Thus, the remaining interpolation nodes (after the first upwind point) are chosen based on smoothness considerations. In particular, this approach will, if at all possible, not run an interpolant across a jump in the data. However, very small interpolation overshoots do occur near extrema in the nodal data, as they must, since any smooth function will slightly overshoot its values as sampled at discrete points near extrema. This is the sense in which the method is only *essentially* nonoscillatory.

14.3.2   Constructing the Numerical Flux Function

We define the numerical flux function through the relation

$$f(u_i)_x = \frac{\mathcal{F}_{i+1/2} - \mathcal{F}_{i-1/2}}{\Delta x} \tag{14.13}$$

as in equation (14.12). To obtain a convenient algorithm for computing this numerical flux function, we define $h(x)$ implicitly through the equation

$$f(u(x)) = \frac{1}{\Delta x} \int_{x-\Delta x/2}^{x+\Delta x/2} h(y)\, dy, \tag{14.14}$$

and note that taking a derivative on both sides of this equation yields

$$f(u(x))_x = \frac{h(x + \Delta x/2) - h(x - \Delta x/2)}{\Delta x}, \tag{14.15}$$

which shows that h is identical to the numerical flux function at the cell walls. That is, $\mathcal{F}_{i\pm1/2} = h(x_{i\pm1/2})$ for all i. We calculate h by finding its primitive

$$H(x) = \int_{x_{-1/2}}^{x} h(y)\, dy \tag{14.16}$$

using polynomial interpolation, and then take a derivative to get h. Note that we do not need the zeroth-order divided differences of H that vanish with the derivative.

The zeroth order divided differences, $D^0_{i+1/2}$ and all higher-order even divided differences of H exist at the cell walls and have the subscript $i\pm1/2$. The first order divided differences D^1_i and all higher-order odd divided differences of H exist at the grid points and will have the subscript i. The first-order divided differences of H are

$$D^1_i H = \frac{H(x_{i+1/2}) - H(x_{i-1/2})}{\Delta x} = f(u(x_i)), \tag{14.17}$$

where the second equality sign comes from

$$H(x_{i+1/2}) = \int_{x_{-1/2}}^{x_{i+1/2}} h(y)\, dy = \sum_{j=0}^{i} \left(\int_{x_{j-1/2}}^{x_{j+1/2}} h(y)\, dy \right) \tag{14.18}$$

$$= \triangle x \sum_{j=0}^{i} f(u(x_j)), \tag{14.19}$$

and the higher divided differences are

$$D_{i+1/2}^2 H = \frac{f(u(x_{i+1})) - f(u(x_i))}{2\triangle x} = \frac{1}{2} D_{i+1/2}^1 f, \tag{14.20}$$

$$D_i^3 H = \frac{1}{3} D_i^2 f, \tag{14.21}$$

continuing in that manner.

According to the rules of polynomial interpolation, we can take any path along the divided difference table to construct H, although not all paths give good results. ENO reconstruction consists of two important features. First, choose $D_i^1 H$ in the upwind direction. Second, choose higher-order divided differences by taking the smaller in absolute value of the two possible choices. Once we construct $H(x)$, we evaluate $H'(x_{i+1/2})$ to get the numerical flux $\mathcal{F}_{i+1/2}$.

14.3.3   ENO-Roe Discretization (Third-Order Accurate)

For a specific cell wall, located at $x_{i_0+1/2}$, we find the associated numerical flux function $\mathcal{F}_{i_0+1/2}$ as follows. First, we define a characteristic speed $\lambda_{i_0+1/2} = f'(u_{i_0+1/2})$, where $u_{i_0+1/2} = (u_{i_0} + u_{i_0+1})/2$ is defined using a standard linear average. Then, if $\lambda_{i_0+1/2} > 0$, set $k = i_0$. Otherwise, set $k = i_0 + 1$. Define

$$Q_1(x) = (D_k^1 H)(x - x_{i_0+1/2}). \tag{14.22}$$

If $|D_{k-1/2}^2 H| \le |D_{k+1/2}^2 H|$, then $c = D_{k-1/2}^2 H$ and $k^\star = k - 1$. Otherwise, $c = D_{k+1/2}^2 H$ and $k^\star = k$. Define

$$Q_2(x) = c(x - x_{k-1/2})(x - x_{k+1/2}). \tag{14.23}$$

If $|D_{k^\star}^3 H| \le |D_{k^\star+1}^3 H|$, then $c^\star = D_{k^\star}^3 H$. Otherwise, $c^\star = D_{k^\star+1}^3 H$. Define

$$Q_3(x) = c^\star(x - x_{k^\star-1/2})(x - x_{k^\star+1/2})(x - x_{k^\star+3/2}). \tag{14.24}$$

Then

$$\mathcal{F}_{i_0+1/2} = H'(x_{i_0+1/2}) = Q_1'(x_{i_0+1/2}) + Q_2'(x_{i_0+1/2}) + Q_3'(x_{i_0+1/2}), \tag{14.25}$$

which simplifies to

$$\mathcal{F}_{i_0+1/2} = D_k^1 H + c\left(2(i_0 - k) + 1\right)\triangle x + c^\star\left(3(i_0 - k^\star)^2 - 1\right)(\triangle x)^2. \tag{14.26}$$

14.3.4   ENO-LLF Discretization (and the Entropy Fix)

The ENO-Roe discretization can admit entropy-violating expansion shocks near sonic points. That is, at a place where a characteristic velocity changes sign (a sonic point) it is possible to have a stationary expansion shock solution with a discontinuous jump in value. If this jump were smoothed out even slightly, it would break up into an expansion fan (i.e., rarefaction) and dissipate, which is the desired physical solution. For a specific cell wall $x_{i_0+1/2}$, if there are no nearby sonic points, then we use the ENO-Roe discretization. Otherwise, we add high-order dissipation to our calculation of $\mathcal{F}_{i_0+1/2}$ to break up any entropy-violating expansion shocks. We call this entropy-fixed version of the ENO-Roe discretization ENO-Roe fix or just ENO-RF. More specifically, we use $\lambda_{i_0} = f'(u_{i_0})$ and $\lambda_{i_0+1} = f'(u_{i_0+1})$ to decide whether there are sonic points in the vicinity. If λ_{i_0} and λ_{i_0+1} agree in sign, we use the ENO-Roe discretization where $\lambda_{i_0+1/2}$ is taken to have the same sign as λ_{i_0} and λ_{i_0+1}. Otherwise, we use the ENO-LLF entropy fix discretization given below. Note that ENO-LLF is applied at both expansions where $\lambda_{i_0} < 0$ and $\lambda_{i_0+1} > 0$ and at shocks where $\lambda_{i_0} > 0$ and $\lambda_{i_0+1} < 0$. While this adds extra numerical dissipation at shocks, it is not harmful, since shocks are self-sharpening. In fact, this extra dissipation provides some viscous regularization which is especially desirable in multiple spatial dimensions. For this reason, authors sometimes use the ENO-LLF method everywhere as opposed to mixing in ENO-Roe discretizations where the upwind direction is well determined by the eigenvalues λ.

The ENO-LLF discretization is formulated as follows. Consider two primitive functions H^+ and H^-. We compute a divided difference table for each of them, with their first divided differences being

$$D_i^1 H^\pm = \frac{1}{2}f(u_i) \pm \frac{1}{2}\alpha_{i_0+1/2}u_i, \qquad (14.27)$$

where

$$\alpha_{i_0+1/2} = \max\left(|\lambda_{i_0}|, |\lambda_{i_0+1}|\right) \qquad (14.28)$$

is our dissipation coefficient. The second and third divided differences, $D_{i+1/2}^2 H^\pm$ and $D_i^3 H^\pm$, are then defined in the standard way, like those of H.

For H^+, set $k = i_0$. Then, replacing H with H^+ everywhere, define $Q_1(x)$, $Q_2(x)$, $Q_3(x)$, and finally $\mathcal{F}_{i_0+1/2}^+$ using the ENO-Roe algorithm above. For H^-, set $k = i_0 + 1$. Then, replacing H with H^- everywhere, define $Q_1(x)$, $Q_2(x)$, $Q_3(x)$, and finally $\mathcal{F}_{i_0+1/2}^-$ again by using the ENO-Roe algorithm above. Finally,

$$\mathcal{F}_{i_0+1/2} = \mathcal{F}_{i_0+1/2}^+ + \mathcal{F}_{i_0+1/2}^- \qquad (14.29)$$

is the new numerical flux function with added high-order dissipation.

14.4   Multiple Spatial Dimensions

In multiple spatial dimensions, the ENO discretization is applied independently using a dimension-by-dimension discretization. For example, consider a two-dimensional conservation law

$$u_t + f(u)_x + g(u)_y = 0 \qquad (14.30)$$

on a rectangular 2D grid. Here, we sweep through the grid from bottom to top performing ENO on 1D horizontal rows of grid points to evaluate the $f(u)_x$ term. The $g(u)_y$ term is evaluated in a similar manner, sweeping through the grid from left to right performing ENO on 1D vertical rows of grid points. Once we have a numerical approximation to each of the spatial terms, we update the entire equation in time with a method-of-lines approach using, for example, a TVD Runge-Kutta method.

14.5   Systems of Conservation Laws

In general, a hyperbolic system will simultaneously contain a mixture of processes: smooth bulk convection and wave motion, and discontinuous processes involving contacts, shocks, and rarefactions. For example, if a gas in a tube is initially prepared with a jump in the states (density, velocity, and temperature) across some surface, as the evolution proceeds in time these jumps will break up into a combination of shocks, rarefactions, and contacts, in addition to any bulk motion and sound waves that may exist or develop.

The hyperbolic systems we encounter in physical problems are written in what are effectively the mixed variables where the apparent behavior is quite complicated. A transformation is required to decouple them back into unmixed fields that exhibit the pure contact, shock, and rarefaction phenomena (as well as bulk convection and waves). In a real system, this perfect decoupling is not possible, because the mixing is nonlinear, but it can be achieved approximately over a small space and time region, and this provides the basis for the theoretical understanding of the structure of general hyperbolic systems of conservation laws. This is called a transformation to characteristic variables. As we shall see, this transformation also provides the basis for designing appropriate numerical methods.

Consider a simple hyperbolic system of N equations

$$\vec{U}_t + [\vec{F}(\vec{U})]_x = 0 \qquad (14.31)$$

in one spatial dimension. The basic idea of characteristic numerical schemes is to transform this nonlinear system to a system of N (nearly) independent scalar equations of the form

$$u_t + \lambda u_x = 0 \qquad (14.32)$$

and discretize each scalar equation independently in an upwind biased fashion using the characteristic velocity λ. Then transform the discretized system back into the original variables.

14.5.1 The Eigensystem

In a smooth region of the flow, we can get a better understanding of the structure of the system by expanding out the derivative as

$$\vec{U}_t + J\vec{U}_x = 0, \tag{14.33}$$

where $J = \partial \vec{F}/\partial \vec{U}$ is the Jacobian matrix of the convective flux function. If J were a diagonal matrix with real diagonal elements, this system would decouple into N independent scalar equations, as desired. In general, J is not of this form, but we can transform this system to that form by multiplying through by a matrix that diagonalizes J. If the system is indeed hyperbolic, J will have N real eigenvalues $\lambda^p, p = 1, \ldots, N$, and N linearly independent right eigenvectors. If we use these as columns of a matrix R, this is expressed by the matrix equation

$$JR = R\Lambda, \tag{14.34}$$

where Λ is a diagonal matrix with the elements $\lambda^p, p = 1, \ldots, N$, on the diagonal. Similarly, there are N linearly independent left eigenvectors. When these are used as the rows of a matrix L, this is expressed by the matrix equation

$$LJ = \Lambda L, \tag{14.35}$$

where L and R can be chosen to be inverses of each other, $LR = RL = I$. These matrices transform to a system of coordinates in which J is diagonalized,

$$LJR = \Lambda, \tag{14.36}$$

as desired.

Suppose we want to discretize our equation at the node x_0, where L and R have values L_0 and R_0. To get a locally diagonalized form, we multiply our system equation by the constant matrix L_0 that nearly diagonalizes J over the region near x_0. We require a constant matrix so that we can move it inside all derivatives to obtain

$$[L_0\vec{U}]_t + L_0JR_0[L_0\vec{U}]_x = 0, \tag{14.37}$$

where we have inserted $I = R_0L_0$ to put the equation in a more recognizable form. The spatially varying matrix L_0JR_0 is exactly diagonalized at the point x_0, with eigenvalues λ_0^p, and it is nearly diagonalized at nearby points. Thus the equations are sufficiently decoupled for us to apply upwind biased discretizations independently to each component, with λ_0^p determining the upwind biased direction for the pth component equation. Once this system is fully discretized, we multiply the entire system by $L_0^{-1} = R_0$ to return to the original variables.

In terms of our original equation (14.31), our procedure for discretizing at a point x_0 is simply to multiply the entire system by the left eigenvector matrix L_0,

$$[L_0\vec{U}]_t + [L_0\vec{F}(\vec{U})]_x = 0, \tag{14.38}$$

and discretize the N scalar components of this system, indexed by p,

$$[(L_0\vec{U})_p]_t + [(L_0\vec{F}(\vec{U}))_p]_x = 0 \tag{14.39}$$

independently, using upwind biased differencing with the upwind direction for the pth equation determined by the sign of λ^p. We then multiply the resulting spatially discretized system of equations by R_0 to recover the spatially discretized fluxes for the original variables

$$\vec{U}_t + R_0\Delta(L_0\vec{F}(\vec{U})) = 0, \tag{14.40}$$

where Δ stands for the upwind biased discretization operator, i.e., either the ENO-RF or ENO-LLF discretization.

We call λ^p the pth characteristic velocity or speed, $(L_0\vec{U})_p = \vec{L}_0^p \cdot \vec{U}$ the pth characteristic state or field (here L^p denotes the pth row of L, i.e., the pth left eigenvector of J), and $(L_0\vec{F}(\vec{U}))_p = \vec{L}_0^p \cdot \vec{F}(\vec{U})$ the pth characteristic flux. According to the local linearization, it is approximately true that the pth characteristic field rigidly translates in space at the pth characteristic velocity. Thus this decomposition corresponds to the local physical propagation of independent waves or signals.

14.5.2   Discretization

At a specific flux location $x_{i_0+1/2}$ midway between two grid nodes, we wish to find the vector numerical flux function $\vec{\mathcal{F}}_{i_0+1/2}$. First we evaluate the eigensystem at the point $x_{i_0+1/2}$ using the standard average $\vec{U}_{i_0+1/2} = (U_i + U_{i+1})/2$. Note that there are more advanced ways to evaluate the eigensystem, as detailed by Donat and Marquina [55]; see also Fedkiw et al. [65]. Then, in the pth characteristic field we have an eigenvalue $\lambda^p(\vec{U}_{i_0+1/2})$, left eigenvector $\vec{L}^p(\vec{U}_{i_0+1/2})$, and right eigenvector $\vec{R}^p(\vec{U}_{i_0+1/2})$. We put $\vec{U}$ values and $\vec{F}(\vec{U})$ values into the pth characteristic field by taking the dot product with the left eigenvector,

$$u = \vec{L}^p(\vec{U}_{i_0+1/2}) \cdot \vec{U} \tag{14.41}$$

$$f(u) = \vec{L}^p(\vec{U}_{i_0+1/2}) \cdot \vec{F}(\vec{U}) \tag{14.42}$$

where u and $f(u)$ are scalars. Once in the characteristic field we perform a scalar version of the conservative ENO scheme, obtaining a scalar numerical flux function $\mathcal{F}_{i_0+1/2}$ in the scalar field. We take this flux out of the characteristic field by multiplying by the right eigenvector,

$$\vec{\mathcal{F}}^p_{i_0+1/2} = \mathcal{F}_{i_0+1/2} \vec{R}^p(\vec{U}_{i_0+1/2}), \qquad (14.43)$$

where $\vec{\mathcal{F}}^p_{i_0+1/2}$ is the portion of the numerical flux function $\vec{\mathcal{F}}_{i_0+1/2}$ from the pth field. Once we have evaluated the contribution to the numerical flux function from each field, we get the total numerical flux by summing the contributions from each field,

$$\vec{\mathcal{F}}_{i_0+1/2} = \sum_p \vec{\mathcal{F}}^p_{i_0+1/2}, \qquad (14.44)$$

completing the evaluation of our numerical flux function at the point $x_{i_0+1/2}$.

14.6   Compressible Flow Equations

The equations for one-phase compressible flow are a general system of convection-diffusion-reaction conservation equations in up to three spatial dimensions. For example, in two spatial dimensions, the equations are of the form

$$\vec{U}_t + \vec{F}(\vec{U})_x + \vec{G}(\vec{U})_y = \vec{F}_d(\nabla\vec{U})_x + \vec{G}_d(\nabla\vec{U})_y + \vec{S}(\vec{U}), \qquad (14.45)$$

where $\vec{U}$ is the vector of conserved variables, $\vec{F}(\vec{U})$ and $\vec{G}(\vec{U})$ are the vectors of convective fluxes, $\vec{F}_d(\nabla\vec{U})$ and $\vec{G}_d(\nabla\vec{U})$ are the vectors of diffusive fluxes, and $\vec{S}(\vec{U})$ is the vector of reaction terms. Again, for high-speed flow with shocks, one can usually ignore the diffuse fluxes. We choose to ignore the source terms (e.g., the effects of chemical reaction) here as well. For more details on the diffuse terms and the source terms, see, for example, Fedkiw et al. [68, 67, 69].

The inviscid Euler equations for one-phase compressible flow in the absence of chemical reactions are then

$$\vec{U}_t + \vec{F}(\vec{U})_x + \vec{G}(\vec{U})_y + \vec{H}(\vec{U})_z = 0, \qquad (14.46)$$

which can be written in detail as

$$\begin{pmatrix} \rho \\ \rho u \\ \rho v \\ \rho w \\ E \end{pmatrix}_t + \begin{pmatrix} \rho u \\ \rho u^2 + p \\ \rho uv \\ \rho uw \\ (E+p)u \end{pmatrix}_x + \begin{pmatrix} \rho v \\ \rho uv \\ \rho v^2 + p \\ \rho vw \\ (E+p)v \end{pmatrix}_y + \begin{pmatrix} \rho w \\ \rho uw \\ \rho vw \\ \rho w^2 + p \\ (E+p)w \end{pmatrix}_z = 0$$

$$(14.47)$$

where ρ is the density, $\vec{V} = \langle u, v, w \rangle$ are the velocities, E is the total energy per unit volume, and p is the pressure. The total energy is the sum of the internal energy and the kinetic energy,

$$E = \rho e + \rho(u^2 + v^2 + w^2)/2, \qquad (14.48)$$

where e is the internal energy per unit mass. The two-dimensional Euler equations are obtained by setting $w = 0$, while the one-dimensional Euler equations are obtained by setting both $v = 0$ and $w = 0$.

The pressure can be written as a function of density and internal energy, $p = p(\rho, e)$. The speed of sound is defined by

$$c = \sqrt{p_\rho + \frac{p p_e}{\rho^2}}, \qquad (14.49)$$

where p_ρ and p_e are partial derivatives of the pressure with respect to the density and internal energy, respectively.

14.6.1 Ideal Gas Equation of State

For an ideal gas we have $p = \rho R T$ where $R = R_u/M$ is the specific gas constant, with $R_u \approx 8.31451$ J/(mol K) the universal gas constant and M the molecular weight of the gas. Also valid for an ideal gas is $c_p - c_v = R$, where c_p is the specific heat at constant pressure and c_v is the specific heat at constant volume. The ratio of specific heats is given by $\gamma = c_p/c_v$. For an ideal gas, one can write

$$de = c_v \, dT, \qquad (14.50)$$

and assuming that c_v does not depend on temperature (calorically perfect gas), integration yields

$$e = e_o + c_v T, \qquad (14.51)$$

where e_o is not uniquely determined, and one could choose any value for e at 0 K (although one needs to use caution when dealing with more than one material to be sure that integration constants are consistent with the heat release in any chemical reactions that occur). For more details, see, e.g., Atkins [7]. Note that

$$p = \rho R T = \frac{R}{c_v} \rho(e - e_o) = (\gamma - 1)\rho(e - e_o), \qquad (14.52)$$

and equation (14.51) are used frequently with $e_o = 0$ arbitrarily for simplicity.

14.6.2 Eigensystem

For brevity we consider only the two-dimensional eigensystem here. The two-dimensional Euler equations can be obtained by setting $w = 0$ so that both the fourth equation in equations (14.47) and the entire $\vec{H}(\vec{U})_z$ term vanish.

The eigenvalues and eigenvectors for the Jacobian matrix of $\vec{F}(\vec{U})$ are obtained by setting $A = 1$ and $B = 0$ in the following formulas, while those for the Jacobian of $\vec{G}(\vec{U})$ are obtained with $A = 0$ and $B = 1$.

The eigenvalues are

$$\lambda^1 = \hat{u} - c, \quad \lambda^2 = \lambda^3 = \hat{u}, \quad \lambda^4 = \hat{u} + c, \tag{14.53}$$

and the eigenvectors are

$$\vec{L}^1 = \left(\frac{b_2}{2} + \frac{\hat{u}}{2c}, -\frac{b_1 u}{2} - \frac{A}{2c}, -\frac{b_1 v}{2} - \frac{B}{2c}, \frac{b_1}{2} \right), \tag{14.54}$$

$$\vec{L}^2 = (1 - b_2, b_1 u, b_1 v, -b_1), \tag{14.55}$$

$$\vec{L}^3 = (\hat{v}, B, -A, 0), \tag{14.56}$$

$$\vec{L}^4 = \left(\frac{b_2}{2} - \frac{\hat{u}}{2c}, -\frac{b_1 u}{2} + \frac{A}{2c}, -\frac{b_1 v}{2} + \frac{B}{2c}, \frac{b_1}{2} \right), \tag{14.57}$$

$$\vec{R}^1 = \begin{pmatrix} 1 \\ u - Ac \\ v - Bc \\ H - \hat{u}c \end{pmatrix}, \quad \vec{R}^2 = \begin{pmatrix} 1 \\ u \\ v \\ H - 1/b_1 \end{pmatrix}, \tag{14.58}$$

$$\vec{R}^3 = \begin{pmatrix} 0 \\ B \\ -A \\ -\hat{v} \end{pmatrix}, \quad \vec{R}^4 = \begin{pmatrix} 1 \\ u + Ac \\ v + Bc \\ H + \hat{u}c \end{pmatrix}, \tag{14.59}$$

where

$$q^2 = u^2 + v^2, \quad \hat{u} = Au + Bv, \quad \hat{v} = Av - Bu, \tag{14.60}$$

$$\Gamma = p_e/\rho, \quad c = \sqrt{p_\rho + \frac{\Gamma p}{\rho}}, \quad H = (E + p)/\rho, \tag{14.61}$$

$$b_1 = \Gamma/c^2, \quad b_2 = 1 + b_1 q^2 - b_1 H. \tag{14.62}$$

The choice of eigenvectors one and four is unique (up to scalar multiples), but the choice for eigenvectors two and three is not unique. Any two independent vectors from the span of eigenvectors two and three could be used instead. In fact, the numerical method designed by Fedkiw et al. [67] exploits this fact.

The eigensystem for the one-dimensional Euler equations is obtained by setting $v = 0$.

14.6.3 Numerical Approach

Since the three-dimensional Euler equations are a system of conservation laws, the methods outlined earlier in this chapter can be applied in a straight-foward fashion. That is, each of $\vec{F}(\vec{U})_x$, $\vec{G}(\vec{U})_y$, and $\vec{H}(\vec{U})_z$ can

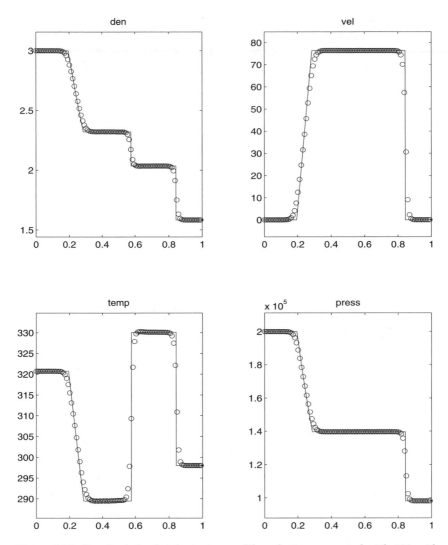

Figure 14.1. Standard shock tube test case. The solution computed with 100 grid cells is depicted by circles, while the exact solution is drawn as a solid line.

be independently approximated using either an ENO-RF or ENO-LLF discretization scheme. A sample calculation in one spatial dimension is shown in Figure 14.1. The initial data consisting of two constant states form (from left to right) a rarefaction wave, a contact discontinuity, and a shock wave. The solution computed with 100 grid cells is depicted by circles, while the exact solution is drawn as a solid line. The figures show solutions at a later time for density, velocity, temperature, and pressure.

15
Two-Phase Compressible Flow

15.1 Introduction

Chronologically, the first attempt to use the level set method for flows involving external physics was in the area of two-phase inviscid compressible flow. Mulder et al. [115] appended the level set equation

$$\phi_t + \vec{V} \cdot \nabla\phi = 0 \qquad (15.1)$$

to the standard equations for one-phase compressible flow, equation (14.47). Here $\vec{V}$ is taken to be the velocity of the compressible flow field, so that the zero level set of ϕ corresponds to particle velocities and can be used to track an interface separating two different compressible fluids. The sign of ϕ is used to identify which gas occupied which region, i.e., to determine the local equation of state. In [115], only gamma law gas models were considered, with $\gamma = \gamma_1$ for $\phi > 0$ and $\gamma = \gamma_2$ for $\phi \leq 0$. Later, Karni [93] pointed out that this method suffered from spurious oscillations at the interface. Figure 15.1 shows a sample calculation using the method proposed in [115]. Here a right going shock wave impinges upon the interface, producing both reflected and transmitted shock waves. Note the spurious oscillations in both the pressure and the velocity profiles near the centrally located contact discontinuity that separates the two different gamma-law gases.

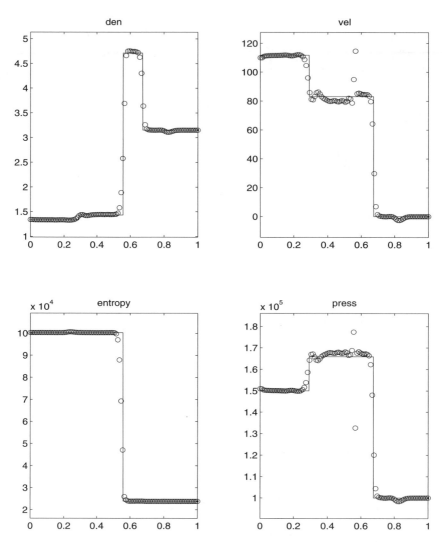

Figure 15.1. Spurious oscillations in pressure and velocity obtained using the method proposed by Mulder et al. [115]. The solution computed with 100 grid cells is depicted by circles, while the exact solution is drawn as a solid line.

15.2   Errors at Discontinuities

The exact solution in Figure 15.1 clearly shows that the pressure and velocity are continuous across the contact discontinuity (in fact, they are constant in this case), while the density and entropy are discontinuous. Since discontinuous quantities indicate the absence of spatial derivatives needed in equation (14.47), one should be suspicious of the behavior of numerical methods in that region. In fact, even the supposedly well-behaved solution shown in Figure 14.1 (page 166) is not entirely adequate near the discontinuities. While the rarefaction wave (to the left) is continuous in nature, both the contact discontinuity and the shock wave should have jump discontinuities in the computed solution. However, the computed solution smears out these discontinuities over a number of grid cells, leading to $O(1)$ errors. Turning our attention to Figure 15.1 we note that the density profile near the contact discontinuity (near the center) should only have values near 1.5 on the left and values near 4.75 on the right; i.e., none of the intermediate values should be present. Intermediate values, such as the one near 2.5, represent a rather significant $O(1)$ error. In light of this, the oscillations in the pressure and velocity shown in Figure 15.1 are no worse than should be expected given the significant errors in the density profile. The only difference is that the density errors are dissipative in nature, while the pressure and velocity errors are dispersive in nature.

Of course, one could argue that dispersive errors are worse than dissipative errors, since dispersive errors produce new extrema, changing the monotonicity of the solution, while dissipative errors only connect two admissible states, producing no new extrema. While this argument is valid for the shock and the contact discontinuity in Figure 14.1 and valid for both of the shocks in Figure 15.1, it is not valid for the contact discontinuity. The gas to the left should never take on values above $\rho = 1.5$, and the gas to the right should never take on values below $\rho = 3$, except at the smeared-out contact discontinuity, which can produce new extrema for the gas to the left and for the gas to the right. Since both of these gases are well-behaved gamma-law gases, it turns out that the oscillations in pressure and velocity can be removed without removing the numerical smearing of the density. However, if one of these gases is replaced with a compressible (and stiff) Tait equation of state for water (that cavitates for densities less than around 999 kg/m^3), it becomes rather difficult to remove the oscillations in the pressure and velocity while still allowing the rather large dissipative errors in the density profile. On the other hand, we will see (below) that it is rather easy to remove all these errors at once.

15.3 Rankine-Hugoniot Jump Conditions

As can be seen in Figure 15.1, the pressure and velocity are continuous across the contact discontinuity, while the density and entropy are discontinuous. If we wish to remove the numerical errors caused by nonphysical smearing of discontinuous quantities, we need to identify exactly what is and what is not continuous across the interface.

In general, conservation of mass, momentum, and energy can be applied to an interface in order to abstract continuous variables. One can place a flux on the interface-oriented tangent to the interface so that material that passes through this flux passes through the interface. If the interface is moving with speed D in the normal direction, this flux will also move with speed D. From conservation, the mass, momentum, and energy that flow into this flux from one side of the interface must flow back out the other side of the interface. Otherwise, there would be a mass, momentum, or energy sink (or source) at the interface, and conservation would be violated. This tells us that the mass, momentum, and energy flux in this moving reference frame (moving at speed D) are continuous variables. We denote the mass, momentum, and energy flux in this moving reference frame by F_ρ, $\vec{F}_{\rho\vec{V}}$, and F_E, respectively. The statement that these variables are continuous is equivalent to the Rankine-Hugoniot jump conditions for an interface moving with speed D in the normal direction.

To define F_ρ, $\vec{F}_{\rho\vec{V}}$, and F_E, we write the equations in conservation form for mass, momentum, and energy as in equation (14.47). The fluxes for these variables are then rewritten in the reference frame of a flux that is tangent to the interface by simply taking the dot product with the normal direction,

$$\left\langle \vec{F}(\vec{U}), \vec{G}(\vec{U}), \vec{H}(\vec{U}) \right\rangle \cdot \vec{N} = \begin{pmatrix} \rho \\ \rho\vec{V}^T \\ E+p \end{pmatrix} V_n + \begin{pmatrix} 0 \\ p\vec{N}^T \\ 0 \end{pmatrix}, \qquad (15.2)$$

where $V_n = \vec{V} \cdot \vec{N}$ is the local fluid velocity normal to the interface, and the superscript "T" designates the transpose operator. Then the measurements are taken in the moving reference frame (speed D) to obtain

$$\begin{pmatrix} \rho \\ \rho\left(\vec{V}^T - D\vec{N}^T\right) \\ \rho e + \frac{\rho|\vec{V} - D\vec{N}|^2}{2} + p \end{pmatrix} (V_n - D) + \begin{pmatrix} 0 \\ p\vec{N}^T \\ 0 \end{pmatrix}, \qquad (15.3)$$

from which we can define

$$F_\rho = \rho(V_n - D), \qquad (15.4)$$

$$\vec{F}_{\rho\vec{V}} = \rho\left(\vec{V}^T - D\vec{N}^T\right)(V_n - D) + p\vec{N}^T, \qquad (15.5)$$

$$F_E = \left(\rho e + \frac{\rho|\vec{V} - D\vec{N}|^2}{2} + p\right)(V_n - D), \qquad (15.6)$$

as continuous variables across the interface. That is, these quantities should be continuous across the interface in order to enforce conservation.

We define the jump in a quantity across the interface as $[\alpha] = \alpha^1 - \alpha^2$, where α^1 is the value of α in fluid 1, and α^2 is the value of α in fluid 2.

Then we can summarize by stating that $[F_\rho] = 0$, $[\vec{F}_{\rho\vec{V}}] = \vec{0}$, and $[F_E] = 0$ across an interface moving with speed D in the normal direction.

15.4   Nonconservative Numerical Methods

Traditionally, Eulerian-based numerical methods for compressible flow are based on the Lax-Wendroff theorem [104], which dictates that numerical methods should be fully conservative, and it is well known that nonconservative methods produce shocks with incorrect speeds and strengths. However, Karni [92] advocated nonconservative form at lower-dimensional (e.g., one-dimensional in a two-dimensional calculation) material interfaces (contact discontinuities) in order to alleviate the oscillations observed in [115]. In [92], full conservation was applied away from interfaces, and a nonconservative method was applied near the interface without adversely affecting the shock speeds or strengths. Since shocks *do not* move at the local interface velocity, any portion of a shock is in contact only with an interface, and thus the nonconservative discretization employed there, on a set of measure zero in space and time, minimizing the accumulation of error.

While it is true that others have used nonconservative discretizations, Karni [92] is responsible for markedly increasing their popularity in the shock-capturing community, where practitioners usually required conservation at all cost. It is interesting to note that many front-tracking and volume-of-fluid schemes are actually nonconservative; i.e., they do not satisfy the strict flux-differencing conservation form usually thought to be required by the Lax-Wendroff theorem. In this sense, many of these schemes share similar properties with the ideology proposed in [92]. For example, consider the front-tracking approach of Pember et al. [129], where a high-order Godunov method is used to obtain a nonconservative update near the tracked interface and a fully conservative update away from the tracked interface. All flow features including shock speeds and strengths as well as the speed of the tracked front are correctly determined, as is ensured by the solutions of the appropriate Riemann problems. Then the authors go one step further and correct the lack of conservation at the interface using a redistribution procedure due to Chern and Colella [45] that is (presumably) not necessary for obtaining a grid-resolved solution, but is used only to maintain exact conservation. In fact, the nature of this redistribution procedure does not allow strict application of the Lax-Wendroff theorem, and one must assume that the correct solutions are obtained because the numerical method is fully conservative except at the lower-dimensional tracked interface, which is updated correctly based on solutions of Riemann problems. Similar loss of exact conservation occurs in volume-of-fluid methods, where nonphysical overshoots may occur in the volume fraction equation;

see Puckett et al. [134]. These overshoots can be ignored when they violate conservation, or redistributed in a manner similar to [45] to preserve exact conservation.

15.5   Capturing Conservation

In summary, conservation of mass, momentum and energy at a discontinuity tells us which variables are continuous across the interface, although as pointed out by Karni [92] one does not necessarily need exact flux-differenced conservative form in order to obtain the correct weak solution. That is, one can instead obtain the correct weak solution in a different manner, for example by solving an associated Riemann problem as in [129]. Therefore, Fedkiw et al. [63] proposed implicitly capturing the Rankine-Hugoniot jump conditions at the interface in order to avoid the intricate details (e.g., multidimensional solution of Riemann problems) that are involved in explicitly enforcing the conservation at the interface. This idea of implicitly capturing conservation as opposed to explicitly enforcing it is in the flavor of level set methods that implicitly capture the location of an interface as opposed to explicitly tracking it. Similar to level set methods, this implicit capturing method for enforcing conservation gives rise to a simple-to-implement numerical algorithm.

This implicit approach to capturing conservation is applied by creating a set of fictitious ghost cells on each side of the interface corresponding to the real fluid on the other side. These ghost cells are populated with a specially chosen (ghost) fluid that implicitly captures the Rankine-Hugoniot jump conditions across the interface. In [63], this method was referred to as the ghost fluid method (GFM).

15.6   A Degree of Freedom

A contact discontinuity (or material interface) has speed $D = V_n$. Rewriting $[F_\rho] = 0$ as $\rho^1(V_n^1 - D) = \rho^2(V_n^2 - D)$ using equation (15.4) and choosing either $D = V_n^1$ or $D = V_n^2$ leads directly to $V_n^1 = V_n^2$ or $[V_n] = 0$. This is our first jump condition, $[V_n] = 0$, implying that the normal velocity is continuous across the interface. Setting $D = V_n^1 = V_n^2$ in $[\vec{F}_{\rho\vec{V}}] = \vec{0}$ leads to $[p]\vec{N}^T = 0$, using equation (15.5) and the fact that the normal is the same on both sides of the interface, i.e., $[\vec{N}] = 0$. Multiplying $[p]\vec{N}^T = 0$ by $\vec{N}$ leads to our second jump condition, $[p] = 0$, implying that the pressure is continuous across the interface. Note that multiplying $[p]\vec{N}^T = 0$ by any tangent vector (there are two of these in three spatial dimensions) leads to $0 = 0$ and the failure to produce a jump condition or a continuous variable. Plugging $D = V_n^1 = V_n^2$ into $[F_E] = 0$ also leads to $0 = 0$ and again a failure to produce a jump condition or a continuous variable. Thus, at a contact discontinuity in three spatial dimensions we have two jump conditions, $[V_n] = 0$ and $[p] = 0$, along with three degrees of freedom corresponding to the $0 = 0$ trivially satisfied jump conditions.

There are five eigenvalues present in the equations of compressible flow in three spatial dimensions. Two of these correspond to the genuinely nonlinear field associated with sound waves, and the other three of these correspond to the linearly degenerate particle velocity. Since a contact discontinuity moves with the linearly degenerate particle velocity, nothing represented by the linearly degenerate fields can cross the interface, meaning that these quantities, the two-dimensional tangential velocity and the entropy, do not cross the interface and may be discontinuous there. In fact, since these quantities are uncoupled across the interface, they are usually discontinuous there. We can write

$$S_t + \vec{V} \cdot \nabla S = 0, \tag{15.7}$$

where S is the entropy, to indicate that entropy is advected along streamlines. Then since S and the contact discontinuity move at the same speed in the normal direction, information associated with S cannot cross the interface. (As a disclaimer we note that equation (15.7) is not valid for streamlines that cross shock waves, i.e., entropy jumps across a shock wave. However, shock waves do not move at the linearly degenerate speed, so this equation is true except for a lower-dimensional subset of space and time.)

Note that in the case of the full viscous Navier-Stokes equations, the physical viscosity imposes continuity of the tangential velocities and thermal conductivity imposes continuity of the temperature.

15.7   Isobaric Fix

In [66], Fedkiw et al. exploited this degree of freedom in order to significantly reduce errors caused by nonphysical wall heating. The well-known "overheating effect" occurs when a shock reflects off of a solid wall boundary, causing overshoots in the temperature and density, while the pressure and velocity remain constant. In one spatial dimension, a solid wall boundary condition can be applied with the aid of ghost cells by constructing a symmetric pressure and density reflection and an asymmetric normal velocity reflection about the solid wall. Then a shock wave impinging on the wall will collide with a shock in the ghost cells that has equal strength traveling in the opposite direction, producing the desired shock reflection. Menikoff [113] and Noh [122] showed that overheating errors are a symptom of smeared-out shock profiles and that sharper shocks usually produce less overheating. They also showed that the pressure and velocity equilibrate quickly, while errors in the temperature and density persist. In order to dissipate these errors in temperature and density, [122] proposed adding artificial heat conduction to the numerical method in a form similar to artificial viscosity. Later, Donat and Marquina [55] proposed a flux-splitting method with a built-in heat conduction mechanism that dissipates these errors throughout the fluid.

For the one-dimensional Euler equations, the Rankine-Hugoniot jump conditions for the solid wall moving at the local flow velocity $D = V_N$ are $[V_N] = 0$, $[p] = 0$, and $0 = 0$. These describe the relationship between the external flow field and the internal one; i.e., both the normal velocity and the pressure must be continuous across the solid wall boundary extending into the ghost cells. Since these jump conditions are inherently part of the equations and thus part of any consistent numerical method, jumps in pressure and velocity are hard to maintain for any duration of time at a solid wall boundary; i.e., jumps between the fluid values and the ghost cell values are quickly dissipated. In this sense, one can think of pressure and velocity equilibration at a solid wall boundary as an intrinsic action of the boundary conditions. There is no such condition for the temperature or the density. In the case of a complete equation of state (see Davis [54]) only one variable in the linearly degenerate field need be defined, and all other variables can be determined from the equation of state relations. In this sense, there is no boundary condition for the linearly degenerate field, as is emphasized by the trivially satisfied jump condition $0 = 0$. Since a solid wall boundary is an initial boundary value problem, the value of the temperature at the wall must come from the initial data, as one can see from equation (15.7) which states that entropy is advected along streamlines of the fluid. This implies that the entropy near the wall stays near the wall, since the wall moves with the local fluid velocity.

In [66], equation (15.7) was used to develop the isobaric fix, which is a boundary condition type of treatment for the linearly degenerate field at a solid wall boundary. The isobaric fix modifies the linearly degenerate field at a solid wall without changing the values of the pressure or the normal velocity. Noting that entropy is advected along streamlines and that the entropy within a fluid is usually continuous, we see that the entropy errors at the wall are repaired using new values of entropy extrapolated from the surrounding flow. For example, replacing the entropy at the wall with the entropy of the neighboring cell gives a first-order accurate value of the entropy at the wall for smooth entropy profiles. Higher-order accurate extrapolation can be used as well, but this has been found to be quite dangerous in practice due to the presence of discontinuous shock waves that can cause large overshoots when one extrapolates with higher than first-order accuracy. In multiple spatial dimensions the solid wall can be represented as the zero isocontour of a level set function and moved rigidly using the level set equation (3.2) where the velocity is chosen as the rigid-wall velocity (or even the velocity of a deforming wall). Then the isobaric fix can be applied using extrapolation of entropy in the normal direction, as discussed in Chapter 8. Note that one does not have to deal with the entropy directly, which can sometimes be difficult to compute for general equations of state, but one can choose any variable corresponding to the equation of state degree of freedom in the linearly degenerate field, e.g., density or temperature.

Although the overheating effect is traditionally discussed in the context of shock reflection from stationary walls, more significant cumulative overheating effects are generated by moving walls. Figure 15.2 shows the errors generated in density and temperature by a solid wall moving from left to right. The wall, initially at rest, is instantaneously accelerated to a velocity of 1000 m/s forming the shock wave seen in the figure. The isobaric fix does not completely eliminate all of the density and temperature errors, but does reduce them by at least an order of magnitude, as shown in Figure 15.3.

15.8   Ghost Fluid Method

In [63], Fedkiw et al. pointed out that a two-phase contact discontinuity could be discretized with techniques similar to those used for the solid-wall boundary, except that they are applied twice, i.e., once for each fluid. Conceptually, each grid point corresponds to one fluid or the other, and ghost cells can be defined at every point in the computational domain so that each grid point contains the mass, momentum, and energy for the real fluid that exists at that point (according to the sign of the level set function) and a ghost mass, momentum, and energy for the other fluid that does not really exist at that grid point (the fluid from the other side of the interface). Once the ghost cells are defined, standard one-phase numerical methods can be used on the entire domain for each fluid; i.e., we now have two separate single-fluid problems. After each fluid is advanced in time, the level set function is updated using equation (3.2) to advect the level set with the local fluid velocity $\vec{V}$, and the sign of the level set function is used to determine the appropriate real fluid values at each grid point. While ghost cells are defined everywhere for the sake of exposition, only a band of 3 to 5 ghost cells is actually needed in practice.

Contact discontinuities move at the local fluid velocity, and the Rankine-Hugoniot jump conditions are $[V_N] = 0$, $[p] = 0$, and $0 = 0$ three times. Only the pressure and normal velocities can be determined from the boundary conditions, while the entropy and both tangential velocities remain undetermined. Since certain properties are discontinuous across the interface, one should be careful in applying finite difference methods across the interface, since differencing discontinuous quantities leads erroneously to terms of the form $1/\triangle x$ that increase without bound as the grid is refined. Therefore, the layer of ghost cells should be introduced so that there is continuity with the neighboring fluid that needs to be discretized. For variables that are already continuous across the interface, e.g., pressure and normal ve-

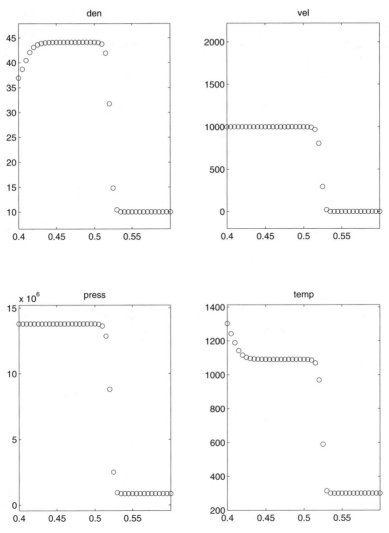

Figure 15.2. Overheating errors in density and temperature generated by a piston moving to the right.

locity, the ghost fluid values can be set equal to the real fluid values at each grid point, implicitly capturing the correct interface values of these variables. This is the key mechanism in coupling the two distinct sets of Euler equations. On the other hand, the discontinuous variables move with the speed of the interface, and information in these variables does not cross the interface and is not coupled to the corresponding information on the other side of the interface. Moreover, in order to avoid numerical smearing or spurious oscillations, these discontinuous variables should not be nonphys-

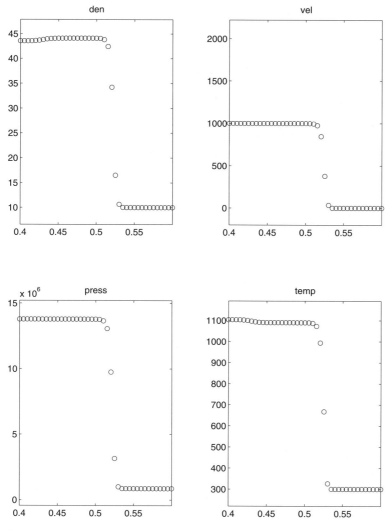

Figure 15.3. The isobaric fix significantly reduces the overheating errors in both density and temperature.

ically coupled together or forced to be continuous across the interface. The most obvious way of defining the discontinuous variables in the ghost cells is by extrapolating that information from the neighboring real fluid nodes; e.g., the entropy can be extrapolated into the ghost cells using extrapolatation in the normal equation in the same fashion as it was in applying the isobaric fix. Again, as with to the isobaric fix, one does not have to deal with the entropy directly, but can choose any variable in the linearly degenerate field, e.g., density or temperature.

In order to to extrapolate the tangential velocity, we first extrapolate the entire velocity field $\vec{V}$. Then, at every cell in the ghost region we have two separate velocity fields, one from the real fluid and one from the extrapolated fluid. For each velocity field, the normal component of velocity, V_N, is put into a vector of length three, $V_N \vec{N}$, and then the tangential velocity field is defined by another vector of length three, $\vec{V} - V_N \vec{N}$. We add together the normal component of velocity, $V_N \vec{N}$, from the real fluid and the tangential component of velocity, $\vec{V} - V_N \vec{N}$, from the extrapolated fluid. This new velocity is our ghost fluid velocity. For the full viscous Navier-Stokes equations, the physical viscosity imposes continuity of the tangential velocities, so that the entire velocity field is continuous. In this case the entire velocity field can be copied over into the ghost cells in a node-by-node fashion. Similar statements hold for the temperature in the presence of a nonzero thermal conductivity.

Next, we describe the one-dimensional method in more detail. Suppose that the interface lies between nodes i and $i+1$. Then fluid 1 is defined to the left and includes node i, while fluid 2 is defined to the right and includes node $i+1$. In order to update fluid 1, we need to define ghost fluid values of fluid 1 at nodes to the right, including node $i+1$. For each of these nodes we define the ghost fluid values by combining fluid 2's pressure and velocity at each node with the entropy of fluid 1 from node i; see Figure 15.4. Likewise, we create a ghost fluid for fluid 2 in the region to the left, including node i. This is done by combining fluid 1's pressure and velocity at each node with the entropy of fluid 2 from node $i + 1$. In order to apply the isobaric fix technique, we change the entropy at node i to be equal to the entropy at node $i - 1$ without modifying the values of the pressure and velocity at node i; see Figure 15.5. Likewise, we change the entropy at node $i + 1$ to be equal to the entropy at node $i + 2$.

An important aspect of this method is its simplicity. We do not need to solve a Riemann problem, consider the Rankine-Hugoniot jump conditions, or solve an initial boundary value problem at the interface. We capture the appropriate interface conditions by defining a fluid that has the pressure and velocity of the real fluid at each point, but the entropy of some other fluid. Consider the case of air and water. In order to solve for the air, we replace the water with *ghost air* that acts like the water in every way (pressure and velocity) but appears to be air (entropy). In order to solve for the water, we replace the air with *ghost water* that acts like the air in every way (pressure and velocity) but appears to be water (entropy). Since the ghost fluids behave in a fashion consistent with the real fluids that they are replacing, the appropriate boundary conditions are captured. Since the ghost fluids have the same entropy as the real fluid that is not replaced, we are solving a one-phase problem.

Figure 15.6 shows how the ghost fluid method removes the spurious osciallations in the pressure and velocity obtained using the method proposed in [115] as shown in Figure 15.1. Note that the density profile remains sharp

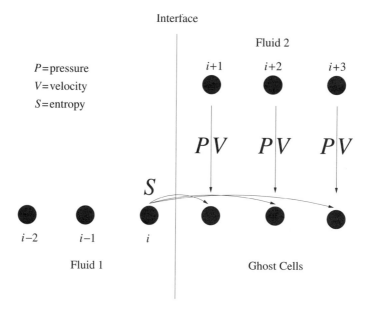

Figure 15.4. Fluid 1's ghost fluid values are constructed by combining fluid 2's pressure and velocity with the entropy of fluid 1 from node i.

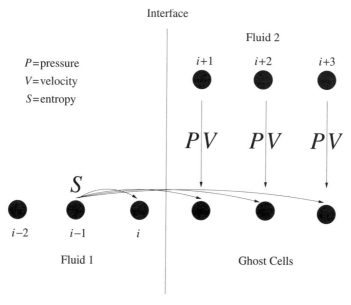

Figure 15.5. In applying the isobaric fix in conjunction with the ghost fluid method, the entropy from node $i - 1$ is used instead of the entropy from node i.

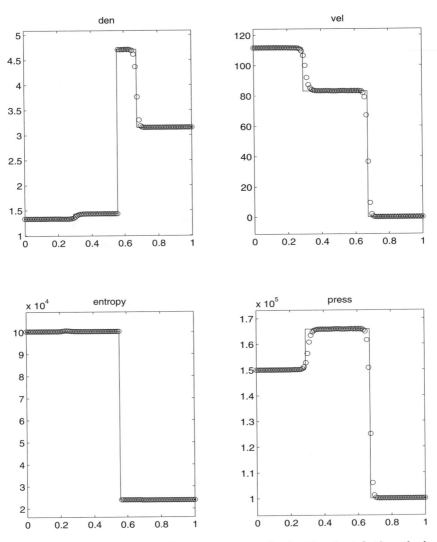

Figure 15.6. The spurious oscillations are removed using the ghost fluid method. Moreover, the density profie remains sharp at the contact discontinuity. The solution computed with 100 grid cells is depicted by circles, while the exact solution is drawn as a solid line.

across the contact discontinuity. While Figure 15.6 is computed using only 100 grid cells, Figure 15.7 is computed with 400 grid cells to illustrate the behavior of the method under mesh refinement. A two-dimensional example of an air shock hitting a helium bubble is shown in Figure 15.8. The black circle indicates the initial location of the Helium bubble before it was hit by the air shock. Figures 15.9, 15.10, 15.11, and 15.12 show two phase-flow calculations where one phase is a gamma-law gas model of air with a

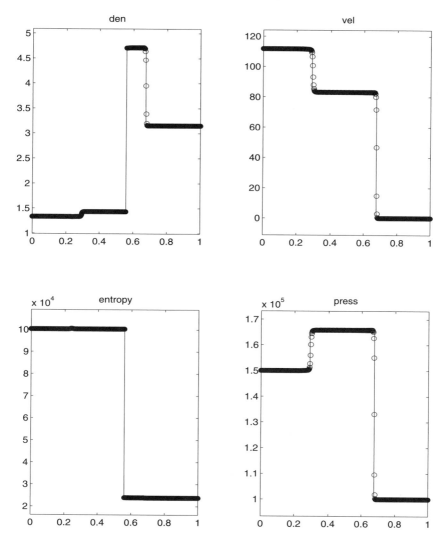

Figure 15.7. This figure illustrates how the ghost fluid method converges as the grid is refined. The solution computed with 400 grid cells is depicted by circles, while the exact solution is drawn as a solid line.

density around 1 kg/m^3 and the other phase is a stiff Tait equation of state model for water with a density around 1000 kg/m^3. In the figures the air is depicted in red and the water is depicted in green. Note that there is no numerical smearing of the density at the interface itself which is fortunate, since water cavitates when it drops to a density slightly above 999 kg/m^3, leading to host of nonphysical problems near the interface. Note that the pressure and velocity are continuous across the interface, although there are kinks in both of these quantities.

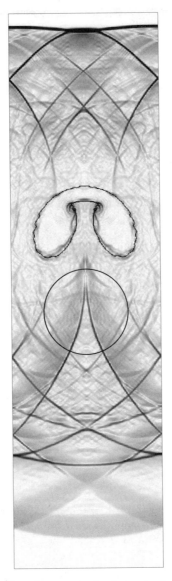

Figure 15.8. Schlieren image for an air shock impinging upon a helium bubble using the ghost fluid method to resolve the contact discontinuity. The black circle indicates the initial location of the helium bubble. This image was generated by Tariq Aslam, of Los Alamos National Laboratory.

15.9 A Robust Alternative Interpolation

The interface values of pressure and normal velocity need to be determined using some sort of interpolation technique, where we note that these variables are continuous but may possess kinks due to differing equations of state across the interface. Copying these variables into the ghost cells in a node-by-node fashion, as proposed above (and in [63]) corresponds to one choice of interpolation. Using the fluid on one side of the interface to determine the interface pressure and the fluid on the other side of the interface to determine the interface normal velocity corresponds to another choice. Different interpolation techniques lead to $O(\triangle x)$ differences in the interface values of pressure and normal velocity, which vanish as the mesh is refined, guaranteeing convergence as the Rankine-Hugoniot jump conditions are implicitly captured.

It is not clear exactly which interpolation technique should be used, and the answer is most likely problem related. For smooth well-behaved problems with commensurate equations of state, the method proposed above (and in [63]) is probably superior, while using one fluid to define the pressure and the other fluid to define the normal velocity is probably superior when one fluid is very stiff compared to the other. For example, consider interactions between a stiff Tait equation-of-state for water and a gamma-law gas model for air as shown, for example, in Figures 15.9, 15.10, 15.11, and 15.12. Since the technique discussed in the last section gives equal weighting to the values of the pressure and normal velocity on both sides of the interface, any kinks in these values will be smeared out to some extent, causing small errors in the captured interface values of these variables. Small errors in the normal velocity of the water create small density errors when the equation for conservation of mass in updated. In turn, these small density errors can lead to large spurious pressure oscillations in the water, since the Tait equation of state is quite stiff. While small errors in the velocity of the air cause the same small density errors, these have little effect on the gas, since the gamma-law gas equation of state is rather robust. Again, since the Tait equation of state is rather stiff, one can expect large variations in the pressure of the water near the interface, which in turn can lead to poor predictions of the interface pressure. While these errors in the interface pressure have a relatively small effect on the heavier water, they can have a rather large effect on the lighter gas. Conversely, since the gamma-law gas equation of state is rather robust, the gas pressure tends to be smooth near the interface and is therefore a good candidate for the interface pressure.

The aforementioned difficulties can be removed in large part by using the water to determine the interface normal velocity and the air to determine the interface pressure, producing a more robust version of the interpolation. When the stiffer fluid (in this case the Tait equation of state water) is updated, pressure is still copied over node by node in the ghost region, while

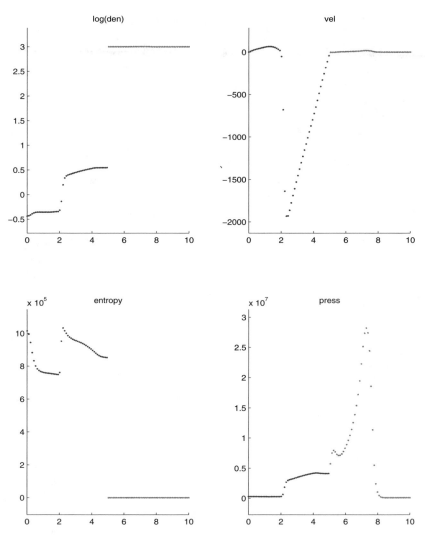

Figure 15.9. The gamma-law gas is depicted in red, while the stiff Tait equation of state water is depicted in green. Note that the log of the density is shown, since the density ratio is approximately 1000 to 1. This calculation uses only 100 grid cells. (See also color figure, Plate 6.)

the total velocity and the entropy are extrapolated into the ghost cells. In updating the fluid with the more robust equation of state (in this case the gamma-law gas air), the normal velocity is still copied over node by node in the ghost region, while the pressure, entropy, and tangential velocity are extrapolated into the ghost cells. This new robust interpolation technique was first proposed by Fedkiw in [62]. Numerical results have shown that

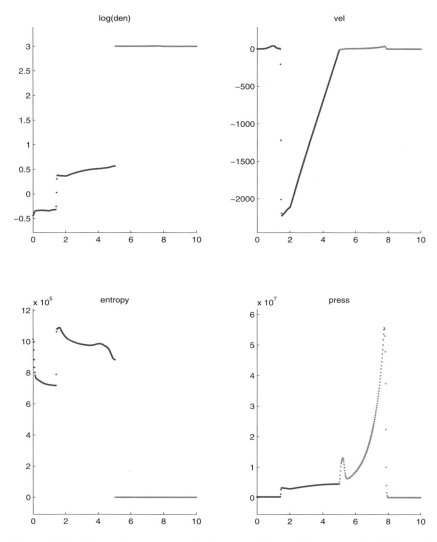

Figure 15.10. This is the same calculation as in Figure 15.9, except that 500 grid cells are used. (See also color figure, Plate 7.)

this new method behaves in a fashion similar to the original method, except for the increased interface dissipation, which leads to greater stability.

In Figures 15.9 and 15.10 an interface separates gas on the left from water on the right. Solid-wall boundary conditions are enforced on both sides of the domain. Initially, a right-going shock wave is located in the gas, and a left-going shock wave is located in the water. These shock waves propagate toward the interface, producing a complex wave interaction. In the figures one can see reflected shock waves traveling outward near $x = 1$

and $x = 8$. The robustnesss of the new interpolation technique is illustrated by the high quality of the solution obtained with as few as 100 grid cells, as shown in Figure 15.9.

In Figures 15.11 and 15.12 interfaces separate gas on the outside of the domain from water on the inside of the domain. A solid-wall boundary is enforced on the left, and an outflow boundary condition is enforced on the right. Initially, all the fluids are moving to the right at 500m/s causing a rarefaction wave to start at the solid wall on the left. This rarefaction wave propagates to the right, slowing down the fluids. Note that it is much easier to slow down the lighter gas as opposed to the heavier water. The figures show the steep pressure profile that forms in the water and acts to to slow the water down. One of the difficulties encountered in [63] was a nonphysical pressure overshoot in the water near the interface on the left. This new robust interpolation technique removes the overshoot, producing a monotone pressure profile near the interface even in the coarse 100-grid-cell solution in Figure 15.11.

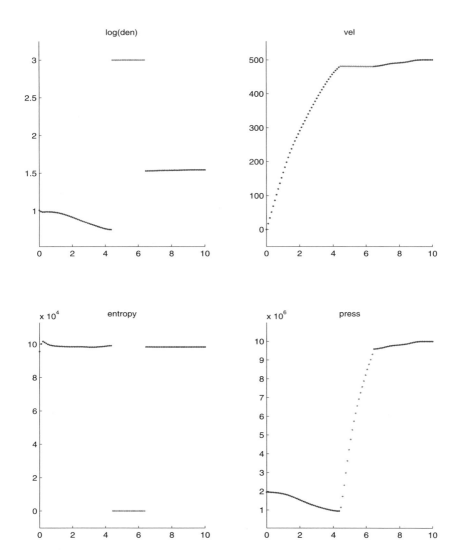

Figure 15.11. In this calculation two interfaces are present, since the air surrounds the water on both sides. This calculation uses only 100 grid cells. (See also color figure, Plate 8.)

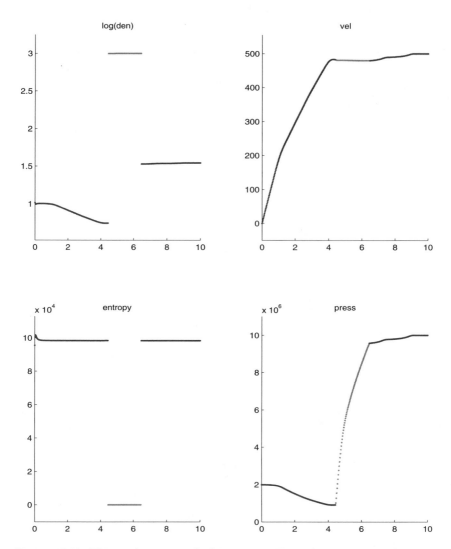

Figure 15.12. This is the same calculation as in Figure 15.11, except that 500 grid cells are used. (See also color figure, Plate 9.)

16
Shocks, Detonations, and Deflagrations

16.1 Introduction

For a contact discontinuity we separated the variables into two sets based on their continuity across the interface. The continuous variables were copied into the ghost fluid in a node-by-node fashion, capturing the correct interface values, while the discontinuous variables were extrapolated in a one-sided fashion to avoid numerical dissipation errors. In order to apply this idea to a general interface moving at speed D in the normal direction, we need to correctly determine the continuous and discontinuous variables across the interface. For example, consider a shock wave where all variables are discontinuous, and extrapolation of all variables for both the preshock and postshock fluids obviously gives the wrong answer, since the physical coupling is ignored. We generally state, *For each degree of freedom that is coupled across a discontinuity, one can define a variable that is continuous across the discontinuity, and all remaining degrees of freedom can be expressed as discontinuous variables that can be extrapolated across the discontinuity in a one-sided fashion*, as the key to extending the GFM. For the Euler equations, conservation of mass, momentum, and energy can be applied to any discontinuity in order to abstract continuous variables; i.e., the Rankine-Hugoniot jump conditions always dictate the coupling between the prediscontinuity and postdiscontinuity fluids. This idea was proposed by Fedkiw et al. [64] to create sharp profiles for shock waves, detonation waves, and deflagration waves.

16.2   Computing the Velocity of the Discontinuity

For a contact discontinuity, we use equation (15.1) to update the interface location, but a more general discontinuity moving at speed D in the normal direction is governed by

$$\phi_t + \vec{W} \cdot \nabla \phi = 0, \tag{16.1}$$

where $\vec{W} = D\vec{N}$ is the velocity of the discontinuity. The authors of [64] proposed a capturing method to compute $\vec{W}$ without explicitly computing it on the interface. Suppose $\vec{U}^{(1)}$ and $\vec{U}^{(2)}$ represent states on different sides of the interface. Then, in general, the velocity of the interface is defined by $\vec{W} = \vec{W}(\vec{U}_{int}^{(1)}, \vec{U}_{int}^{(2)})$ where the "int" subscript designates a variable that has been interpolated to the interface in a one-sided fashion. Generally, $\vec{W}$ is a continuous function of its arguments, and application of $\vec{W} = \vec{W}(\vec{U}^{(1)}, \vec{U}^{(2)})$ in a node-by-node fashion will implicitly capture the correct value of $\vec{W}$ at the interface. This computation of $\vec{W} = \vec{W}(\vec{U}^{(1)}, \vec{U}^{(2)})$ in a node-by-node fashion requires values of $\vec{U}^{(1)}$ and $\vec{U}^{(2)}$ at every node. We accomplish this by extrapolating $\vec{U}^{(1)}$ across the interface into the region occupied by $\vec{U}^{(2)}$ (using equation (8.1)), and likewise extrapolating $\vec{U}^{(2)}$ across the interface into the region occupied by $\vec{U}^{(1)}$. Of course, we only need to extrapolate values in a thin band of grid cells near the interface. When $\vec{W} = D\vec{N}$, i.e., $\vec{W} = \vec{W}(\vec{U}_{int}^{(1)}, \vec{U}_{int}^{(2)}) = \vec{D}(\vec{U}_{int}^{(1)}, \vec{U}_{int}^{(2)})\vec{N}$, we can compute D in a node-by-node fashion and obtain $\vec{W}$ by multiplying D by $\vec{N}$.

When using the ghost fluid method for general discontinuities, we need to accurately determine the interface speed D. For shock waves and detonation waves, D can be found by solving an appropriate Riemann problem in a node-by-node fashion [64]. In fact, there is no reason one cannot solve a Riemann problem in the case of a contact discontinuity as well, using $\vec{W} = D\vec{N}$ in equation (16.1) to update the level set function as opposed to using equation (15.1) with the less-accurate local fluid velocity. In fact, a combination of ghost cells and Riemann problems is commonly used in front tracking algorithms; see, e.g., Glimm et al. [73, 72], where a Riemann problem is solved at the interface and the results are extrapolated into ghost cells. The difference between the ghost fluid method and typical front-tracking algorithms is in the order of operations. Front-tracking algorithms first solve a Riemann problem using flow variables interpolated to the (possibly) multidimensional interface and then extrapolate the results into ghost cells, while the ghost fluid method first extrapolates into ghost cells and then solves the Riemann problem in a node-by-node fashion, removing complications due to interface geometry.

For a deflagration wave, the Riemann problem is not well posed unless the speed of the deflagration (D) is given. However, the G-equation for flame discontinuities, see Markstein [110], represents a flame front as a discontinuity in the same fashion as the level set method, so that one can easily consult the abundant literature on the G-equation to obtain deflagration wave speeds.

16.3 Limitations of the Level Set Representation

The level set function is designed to represent interfaces where the interface crosses material at most once due to an entropy condition; see Sethian [147] and Osher and Seian [126]. Contact discontinuities move with the local material velocity, and thus never cross over material. On the other hand, if one material is being converted into another, then the interface may include a regression rate for this conversion. When the regression rate is based on some sort of chemical reaction, the interface can pass over a material exactly once, changing it into another material. The same chemical reaction cannot occur to a material more than once, and the reverse reaction is usually not physically plausible due to an entropy condition. However, for readily reversible chemical reactions, the zero level set may pass over a material in one direction (the reaction) and then pass back over the same material in the opposite direction (the reverse reaction).

Shocks can be interpreted as the conversion of an uncompressed material into a compressed material. Here D is the shock speed, and the ghost fluid method can be used to follow a lead shock, but since shocks can pass over a material more than once in the same direction, all subsequent shocks must be captured or modeled by separate level set functions.

16.4 Shock Waves

Consider the representation of a lead shock by a level set function where the positive values of ϕ correspond to the unshocked material and the negative values of ϕ correspond to the shocked material. Then the normal $\vec{N}$ points from the shocked material into the unshocked material.

In one spatial dimension, the normal velocity is defined as $V_n = \vec{V} \cdot \vec{N}$, and equations (15.4), (15.5) and (15.6) become

$$F_\rho = \rho(V_n - D) \tag{16.2}$$

$$\vec{F}_{\rho\vec{V}} = \rho(u - D\vec{N}^T)(V_n - D) + p\vec{N}^T \tag{16.3}$$

$$F_E = \left(\rho e + \frac{\rho|u - D\vec{N}|^2}{2} + p \right)(V_n - D) \tag{16.4}$$

where it is useful to define

$$F_{\rho V_n} = \vec{N}\vec{F}_{\rho\vec{V}} = \rho(V_n - D)^2 + p \tag{16.5}$$

and to rewrite equation (16.4) as

$$F_E = \left(\rho e + \frac{\rho(V_n - D)^2}{2} + p\right)(V_n - D) \tag{16.6}$$

using the fact that $\vec{N} = \pm 1$ in one spatial dimension.

Our goal is to implicitly capture the Rankine-Hugoniot jump conditions by implicitly enforcing continuity of F_ρ, $F_{\rho V_n}$, and F_E. These quantities can be evaluated at each real grid node by plugging the local values of the conserved variables into equations (16.2), (16.5) and (16.6) to obtain F_ρ^R, $F_{\rho V_n}^R$, and F_E^R, respectively, where the "R" superscript designates a real grid node value. In order to implicitly capture the values of these variables with ghost node values, we want the ghost node values to result exactly in F_ρ^R, $F_{\rho V_n}^R$, and F_E^R when plugged into equations (16.2), (16.5) and (16.6), That is, we want ghost node values of density, velocity, internal energy, and pressure such that

$$\rho^G(V_n^G - D) = F_\rho^R, \tag{16.7}$$

$$\rho^G(V_n^G - D)^2 + p^G = F_{\rho V_n}^R, \tag{16.8}$$

$$\left(\rho^G e^G + \frac{\rho^G(V_n^G - D)^2}{2} + p^G\right)(V_n^G - D) = F_E^R, \tag{16.9}$$

at each grid node, where the "G" subscript designates a ghost node value. Adding the equation of state for the ghost fluid as

$$p^G = (\gamma^G - 1)\rho^G e^G \tag{16.10}$$

yields four equations for four unknowns, which can be arranged into a quadratic equation for $V_n^G - D$, where

$$V_n^G - D = \frac{\gamma^G F_{\rho V_n}^R}{(\gamma^G + 1)F_\rho^R} \pm \sqrt{\left(\frac{\gamma^G F_{\rho V_n}^R}{(\gamma^G + 1)F_\rho^R}\right)^2 - \frac{2(\gamma^G - 1)F_E^R}{(\gamma^G + 1)F_\rho^R}} \tag{16.11}$$

expresses the two solutions. Choosing one of these two solutions for V_n^G allows us to obtain ρ^G from equation (16.7), p^G from equation (16.8), and e^G from equation (16.10). In addition, $u^G = V_n^G \vec{N}$.

In order to choose the correct solution (of the two choices) from equation (16.11), we have to determine whether the ghost fluid is an unshocked (preshock) fluid or a shocked (postshock) fluid. Node by node, we use the real values of the unshocked fluid to create a shocked ghost fluid. Likewise, we use the real values of the shocked fluid to create an unshocked ghost fluid. If the ghost fluid is a shocked fluid, then D is subsonic relative to the flow; i.e., $V_n^G - c^G < D < V_n^G + c^G$ or $|V_n^G - D| < c^G$. If the ghost fluid is an unshocked fluid, then D is supersonic relative to the flow; i.e., $|V_n^G - D| > c^G$. Therefore, the "$\pm$" sign in equation (16.11) should be chosen to give the minimum value of $|V_n^G - D|$ when a shocked ghost fluid is constructed and the maximum value of $|V_n^G - D|$ for an unshocked ghost fluid.

For a simple nonreacting shock, the shock speed D can be defined directly from the mass balance equation as

$$D = \frac{\rho^{(1)}u^{(1)} - \rho^{(2)}u^{(2)}}{\rho^{(1)} - \rho^{(2)}} \tag{16.12}$$

in a node-by-node fashion. However, this simple definition of the shock speed will erroneously give $D = 0$ in the case of a standard shock tube problem where both fluids are initially at rest. A somewhat better estimate of the shock speed can be derived by combining equation (16.12) with the momentum balance equation to obtain

$$D = \sqrt{\frac{\rho^{(1)}\left(u^{(1)}\right)^2 + p^{(1)} - \rho^{(2)}\left(u^{(2)}\right)^2 - p^{(2)}}{\rho^{(1)} - \rho^{(2)}}}, \tag{16.13}$$

where the shock speed is now dependent on the pressure as well. Note that equations (16.12) and (16.13) are only approximations of D. Clearly, these approximations will lead to nonphysical values of D in certain situations. In fact, D could be infinite or even imaginary. A more robust, but still approximate, value for D can be obtained by evaluating $D = V_n + c$ with the Roe average of $\vec{U}^{(1)}$ and $\vec{U}^{(2)}$ (see, for example, LeVeque [105]), since this is the exact shock speed for an isolated shock wave and never becomes ill-defined. Of course, the best definition of the shock speed can be derived by solving the Riemann problem for the states $\vec{U}^{(1)}$ and $\vec{U}^{(2)}$, although this generally requires an iterative procedure. The interested reader is referred to the ongoing work of Aslam [9] for more details.

Figure 16.1 depicts a standard shock tube test case that was computed using the level set method to track the location of the shock wave and the ghost fluid method to accurately capture the boundary conditions across that shock wave. Note the sharp (nonsmeared) representation of the shock wave.

16.5 Detonation Waves

Strong detonations and Chapman-Jouguet detonations can be approximated as reacting shocks under the assumption that the reaction zone has negligible thickness. Again, assume that $\vec{N}$ points from the reacted material into the unreacted material.

Equations (16.7), (16.8) and (16.9) are still valid, while equation (16.10) becomes

$$p^G = (\gamma^G - 1)\rho^G(e^G - e_o^G), \tag{16.14}$$

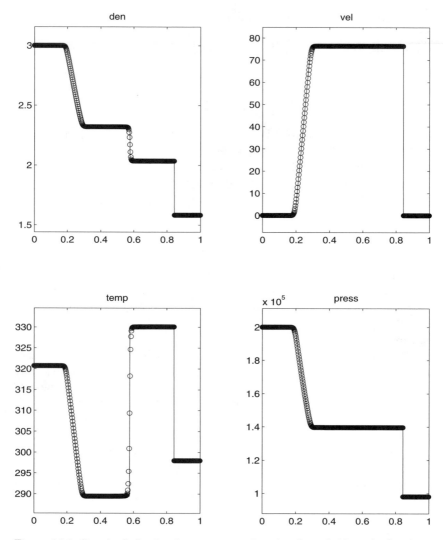

Figure 16.1. Standard shock-tube test case using the ghost fluid method to keep the shock wave sharp. The computed solution is depicted by circles, while the exact solution is drawn as a solid line.

where one can no longer set $e_o = 0$ for both fluids. In detonations, the jump in e_o across the reaction front indicates the energy release in the chemical reaction. Equation 16.11 becomes

$$
V_n^G - D = \frac{\gamma^G F_{\rho V_n}^R}{(\gamma^G + 1) F_\rho^R} \pm \sqrt{\left(\frac{\gamma^G F_{\rho V_n}^R}{(\gamma^G + 1) F_\rho^R}\right)^2 - \frac{2(\gamma^G - 1)}{(\gamma^G + 1)} \left(\frac{F_E^R}{F_\rho^R} - e_o^G\right)}
$$

$$(16.15)$$

where the "±" sign is chosen to give the minimum value of $|V_n^G - D|$ for a reacted ghost fluid and the maximum value of $|V_n^G - D|$ for an unreacted ghost fluid. Equation (16.13) is used for the detonation speed D, although one might want to use a Riemann solver; see, for example, Teng et al. [163].

Figure 16.2 shows an overdriven detonation wave traveling from left to right. A solid-wall boundary condition is enforced on the left, creating a rarefaction wave that will eventually catch up with the overdriven detonation and weaken it to a Chapman-Jouguet detonation. The circles depict the pressure profile calculated with 200 grid cells, while the solid line depicts the computed profile with 800 grid cells. Note that there is no numerical smearing of the leading wave front, which is extremely important when one attempts to eliminate spurious wave speeds for stiff source terms on coarse grids; see, for example, Colella et al. [50].

16.6   Deflagration Waves

For a deflagration wave, equations (16.7), (16.8), (16.9) and (16.14) are used with the jump in e_o equal to the energy release in the chemical reaction. Equation (16.15) is used as well. However, since a deflagration is subsonic, the "±" sign is chosen to give the minimum value of $|V_N^G - D|$ for both the reacted and the unreacted ghost fluids.

For a deflagration, the Riemann problem is not well posed unless the speed of the deflagration is given. Luckily, there is abundant literature on the G-equation for flame discontinuities, and one can consult this literature to obtain appropriate deflagration speeds. For example, using a deflagration velocity from Mulpuru and Wilkin [116],

$$D = V_N + 18.5 \left(\frac{p}{101000Pa}\right)^{.1} \left(\frac{T}{298K}\right)^{1.721} \text{ m/s} \qquad (16.16)$$

evaluated with the velocity, pressure, and temperature of the unreacted gas, we compute the shock deflagration interaction shown in Figure 16.3. Here a left-going shock wave intersects a right-going deflagration wave (unreacted gas to the right), resulting in four waves: a shock, contact, deflagration, and rarefaction from left to right. All the waves are captured, except the deflagration wave, which is tracked with the level set function and resolved with the ghost fluid method. The circles depict the computed solution, while the solid line depicts the exact solution. Note that the pressure drops slightly across a deflagration wave, as opposed to the pressure rise across shock and detonation waves.

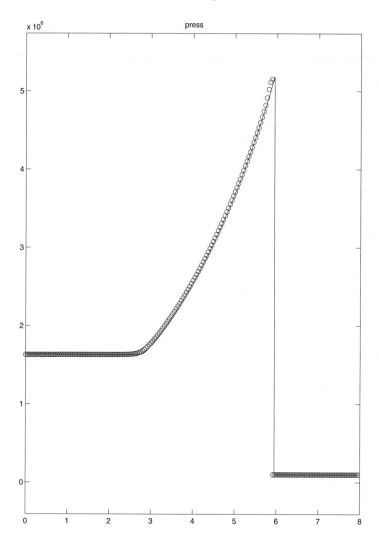

Figure 16.2. Overdriven detonation wave traveling from left to right with a solid-wall boundary condition on the left. The circles depict the pressure profile calculated with 200 grid cells.

16.7 Multiple Spatial Dimensions

In multiple spatial dimensions we use equations (16.2), (16.5), and (16.14) along with

$$\vec{F}_{\rho V_T} = \frac{\vec{F}_{\rho}\vec{V} - F_{\rho V_n}\vec{N}^T}{F_\rho} = \vec{V}^T - V_n\vec{N}^T, \qquad (16.17)$$

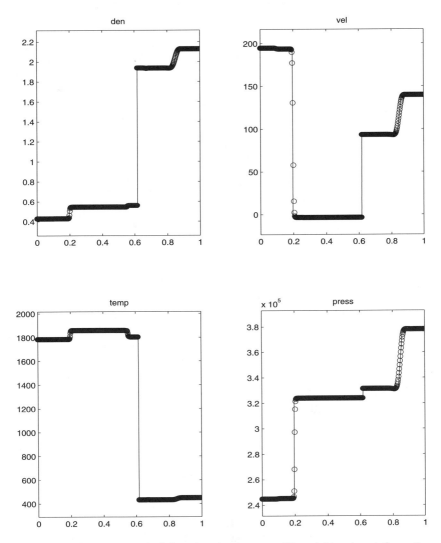

Figure 16.3. Interaction of a left-going shock wave with a right-going deflagration wave, producing four waves: a shock, contact, deflagration, and rarefaction from left to right. The circles depict the computed solution using 400 grid cells, while the solid line depicts the exact solution.

which is valid when $V_n \neq D$, i.e., except for the case of a contact discontinuity. The necessary continuity of this expression implies the well-known fact that tangential velocities are continuous across shock, detonation, and deflagration waves. Note that tangential velocities are not continuous across contact discontinuities unless viscosity is present.

Combining

$$|\vec{V} - D\vec{N}|^2 = |\vec{V}|^2 - 2DV_n + D^2 = |\vec{V}|^2 - V_n^2 + (V_n - D)^2 \qquad (16.18)$$

with

$$|\vec{V}|^2 = V_n^2 + V_{T_1}^2 + V_{T_2}^2, \qquad (16.19)$$

where V_{T_1} and V_{T_2} are the velocities in the tangent directions T_1 and T_2, yields

$$|\vec{V} - D\vec{N}|^2 = V_{T_1}^2 + V_{T_2}^2 + (V_n - D)^2, \qquad (16.20)$$

which can plugged into equation (15.6) to obtain

$$F_E = \left(\rho e + \frac{\rho(V_{T_1}^2 + V_{T_2}^2)}{2} + \frac{\rho(V_n - D)^2}{2} + p \right) (V_n - D) \qquad (16.21)$$

as a rewritten version of equation (15.6). We then write

$$\hat{F}_E = F_E - \frac{F_\rho(V_{T_1}^2 + V_{T_2}^2)}{2} = \left(\rho e + \frac{\rho(V_n - D)^2}{2} + p \right) (V_n - D), \qquad (16.22)$$

which (not coincidently) has the same right-hand side as equation (16.6). In fact, we eventually derive equation (16.15) again, except with F_E^R replaced by the identical $\hat{F}_E^R$.

The main difference between one spatial dimension and multiple spatial dimensions occurs in the treatment of the velocity. The ghost cell velocity $\vec{V}^G$ is obtained by combining the normal velocity of the ghost fluid with the tangential velocity of the real fluid using

$$\vec{V}^G = V_n^G \vec{N} + \vec{V}^R - V_n^R \vec{N}, \qquad (16.23)$$

where $\vec{V}^R - V_n^R \vec{N}$ is the tangential velocity of the real fluid.

Figure 16.4 shows two initially circular deflagration fronts that have merged into a single front. The light colored region surrounding the deflagration fronts is a precursor shock wave that causes the initially circular deflagration waves to deform as they approach each other. Figure 16.5 shows the smooth level set representation of the deflagration wave.

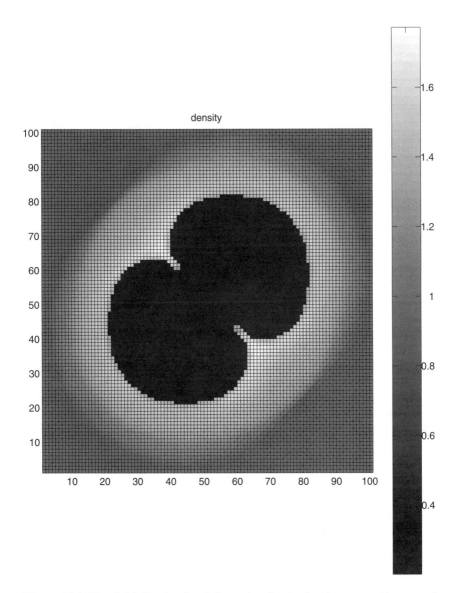

Figure 16.4. Two initially circular deflagration fronts that have recently merged into a single front. The light colored region surrounding the deflagration fronts is a precursor shock wave that causes the initially circular deflagration waves to deform as they approach each other.

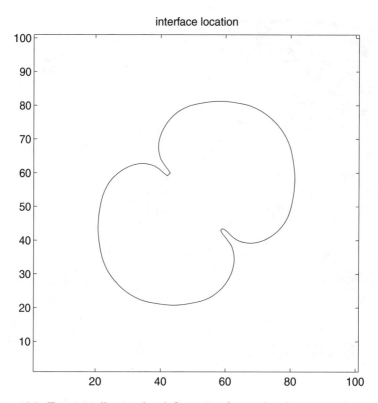

Figure 16.5. Two initially circular deflagration fronts that have recently merged into a single front. Note the smooth level set representation of the deflagration wave.

17

Solid-Fluid Coupling

17.1 Introduction

Solid-fluid interaction problems are still rather difficult for modern computational methods. In general, there are three classical approaches to such problems: One can treat both the solid and the fluid with Eulerian numerical methods, the fluid with an Eulerian numerical method and the solid with a Lagrangian numerical method, or both the solid and the fluid with Lagrangian numerical methods.

Many fluid flows, e.g., high-speed gas flows with strong shocks and large deformations, are difficult to simulate with Lagrangian numerical methods that use artificial viscosity to smear out shock profiles over a number of zones in order to reduce postshock oscillations or ringing. The artificial viscosity can be both problem- and material-dependent. Flows with significant deformations can cause large mesh perturbations and subsequent numerical errors that can be removed only with complicated remeshing and/or mesh generation procedures that tend to be low-order accurate. In particular, vorticity can cause the mesh to tangle and sometimes invert, in which case the calculation needs to be stopped. Eulerian numerical methods intrinsically avoid these mesh-associated difficulties by using a stationary mesh. Furthermore, these schemes capture shocks in a straightforward fashion using conservation and robust limiters, eliminating the need for problem-dependent artificial viscosity formulations. This allows shocks to be modeled with as few as one grid cell without oscillations, whereas Lagrangian numerical methods usually suffer from some amount of postshock oscillations until the shock is spread out over about six grid cells; see, e.g., Benson [14, 15].

While Eulerian numerical methods are superior for high-speed gas flows, they can perform poorly for solid mechanics calculations. The capturing nature of Eulerian methods generally leads to algorithms that are not accurate or robust enough to track time-history variables or material response to loading and damage. On the other hand, Lagrangian numerical methods are extremely accurate and well tested in this area.

Many researchers agree that it is preferable to use Eulerian numerical methods for fluids and Lagrangian numerical methods for solids. Then there are two popular approaches for treating the solid-fluid interface. First, one can smear out the nature of the numerical approximations using a Lagrangian method in the solid and an Eulerian method in the fluid with a "mushy" region in between where the grid moves with an intermediate velocity. That is, the grid velocity is smoothly varying between the Lagrangian mesh velocity and the zero-velocity Eulerian mesh. This is the fundamental idea behind arbitrary Lagrangian-Eulerian (ALE) numerical algorithms; see, e.g., [14]. The problem with this approach is that the variable velocity mesh has not been well studied, and the numerical algorithms employed on it tend to be low-order accurate and suspect. The second approach for treating the solid-fluid interface is to keep the mesh representation sharp so that the Eulerian and Lagrangian meshes are in direct contact. The problem with this approach is that the Lagrangian mesh moves, causing Eulerian mesh points to appear and disappear. In addition, the Eulerian cells tend to have irregular shapes, referred to as cut cells. These cut cells can lead to numerical errors and stiff time-step restrictions; see, e.g., [14] and the references therein, specifically Hancock [78], Noh [121], and McMaster [112], which discuss the PISCES, CEL, and PELE programs, respectively.

In [62], the author took the second approach for the treatment of the solid-fluid interface. However, problems with cut cells were avoided by using ghost cells for the Eulerian mesh. These ghost cells are covered (or partially covered) by the Lagrangian mesh, but are used in the Eulerian finite difference scheme in order to circumvent small time-step restrictions. The ghost cells are defined in a way consistent with a contact discontinuity so that the interface boundary conditions or jump conditions are properly captured. This method also avoids the blending problems associated with covering and uncovering of grid points, since covered real grid nodes are subsequently treated as ghost nodes, and uncovered ghost nodes are subsequently treated as real grid nodes. Moreover, the numerical treatment of the solid-fluid interface does not compromise the solution techniques for the solid or the fluid. That is, once the ghost cells' values are specified, a standard Eulerian program can be used to advance the fluid (and its ghost nodes) in time. A standard Lagrangian program can be used to advance the solid in time as well as to acquire boundary conditions from the Eulerian mesh using both the real grid nodes and the ghost nodes in a standard interpolation procedure. Aivazis et al. [3] successfully used this method to couple an Eulerian hydroprogram to a highly sophisticated Lagrangian materials program designed by Ortiz.

17.2 Lagrange Equations

While there are a number of sophisticated Lagrangian programs, we particularly like the approach of Caramana et al. [24] which allows for implementation of arbitrary forces in a straightforward fashion.

The Lagrange equations are written in nonconservative form with position, velocity, and internal energy as the independent variables. In one spatial dimension, both x and u are defined at the grid nodes, while e is defined at the cell centers located midway between the nodes. To initialize the calculation, the mass of each zone, M^z, is determined, and then the subzonal masses m^z are defined as half the zonal mass. The nodal mass M^p is defined as the sum of the neighboring subzonal masses. The nodal, zonal, and subzonal masses all remain fixed throughout the calculation. At each time step, the location of each grid node is updated according to

$$\frac{x^{n+1} - x^n}{\Delta t} = u^n, \tag{17.1}$$

where Δt is the size of the time step. The velocity at each node is updated using

$$\frac{u^{n+1} - u^n}{\Delta t} = \frac{F^n}{M^p}, \tag{17.2}$$

where F^n is the net force on the grid node. The internal energy in each zone is updated with

$$\frac{e^{n+1} - e^n}{\Delta t} = \frac{H^n}{M^z}, \tag{17.3}$$

where H^n is the heating rate of the zone. One can apply either force or velocity boundary conditions to the grid nodes on the boundary. Velocity boundary conditions are enforced by setting the velocity of a boundary node to the desired boundary velocity instead of solving equation (17.2). Force boundary conditions are applied by adding the boundary force to the net nodal force F^n in equation (17.2).

In two spatial dimensions, both $\vec{X} = \langle x, y \rangle$ and $\vec{V} = \langle u, v \rangle$ are defined at the grid nodes, which are connected in the same fashion as an Eulerian grid, producing quadrilateral zones. Each quadrilateral zone is split into four subzones by connecting the midpoints of opposite edges of the zone. The internal energy is defined at the zone center located at the intersection of the four subzones. To initialize the calculation, the mass of each zone, M^z, is determined, and then the subzonal masses m^z are defined as one-fourth the zonal mass. The nodal mass M^p is defined as the sum of the (at most four) neighboring subzonal masses. Once again, the nodal, zonal, and subzonal masses all remain fixed throughout the calculation. The independent variables are updated with equations (17.1), (17.2) and (17.3) with x, u, and F^n replaced with $\vec{X}$, $\vec{V}$, and $\vec{F}$, respectively. Either force or velocity boundary conditions are applied to the nodes on the boundary.

For three spatial dimensions we refer the reader to Caramana et al. [25].

17.3   Treating the Interface

Boundary conditions need to be imposed on both the Eulerian and Lagrangian grids where the Lagrangian grid partially overlaps the Eulerian grid. First the interface itself needs to be defined, and since the Lagrangian grid nodes move at the local material velocity, these nodes can be used to determine the position of the interface. This interface divides the Eulerian mesh into separate regions, a region populated by real grid nodes and a region populated by ghost nodes. Interface boundary conditions for the Eulerian mesh are imposed by defining mass, momentum, and energy in the ghost nodes. Interface boundary conditions for the Lagrangian mesh are imposed by either specifying the velocity of the grid nodes on the boundary or by specifying the force applied to that boundary.

Since the interface moves with the local material velocity, it can be treated as a contact discontinuity. The pressure and the normal velocity are continuous across the interface, while the entropy and tangential velocities are uncoupled across the interface. The interface values of the uncoupled variables are captured by extrapolating these variables across the interface into the ghost cells. The continuous or coupled variables are determined using the values from both the Eulerian and the Lagrangian meshes.

The interface normal velocity can be determined by applying any number of interpolation techniques to the Eulerian and Lagrangian mesh values. However, one should be careful to define the interface normal velocity in a way that is consistent with the material in the Lagrangian mesh. Perturbations to the velocity of the Lagrangian grid nodes can provide enormous stress due to resistive forces such as material strength. For this reason, in order to determine an accurate (and Lagrangian mesh-consistent) value of the normal velocity at the interface, only the Lagrangian mesh is used to determine the interface velocity, as was done in [78], [112], and [121]. Both calculations use this interface normal velocity, so that $[V_N] = 0$ is enforced. That is, the Lagrangian mesh uses the computed velocities of its boundary nodes, while the Eulerian calculation captures this interface normal velocity by assigning to each ghost node the interface normal velocity of the nearest Lagrangian mesh point on the interface.

Since the interface normal velocity is defined as the velocity of the nodes on the Lagrangian mesh boundary with no contribution from the Eulerian mesh, velocity boundary conditions cannot be enforced on the Lagrangian mesh at the interface. Instead, force boundary conditions are applied by interpolating the Eulerian grid pressure to this Lagrangian interface. In this fashion, the interface pressure is determined using only the Eulerian grid values, ignoring contributions from the Lagrangian mesh, as in [78], [112], and [121]. Both calculations use this interface pressure, so that $[p] = 0$ is enforced. The interface pressure is captured by the Eulerian calculation by extrapolating the pressure across the interface into the ghost cells which is similar to the treatment of entropy and tangential velocity. Then the interface pressure is interpolated from the Eulerian grid in order to apply force boundary conditions to the Lagrangian mesh.

Noh [121] suggested that it might be better to use some average of the Lagrangian and Eulerian grid values for determining the pressure at the interface. For Lagrangian calculations with artificial viscosity and material strength, the jump condition implies that the net stress in the normal direction is continuous, not just the pressure. Therefore, this advocated averaging procedure would need to take place between the pressure in the fluid and the normal component of the net stress in the normal direction in the solid. However, this can be dangerous, for example, when the Lagrangian material is in tension, since near-zero or negative stress might be calculated at the interface. While Lagrangian methods can be quite robust under tension, Eulerian methods can suffer a number of problems in treating near-zero or negative pressures associated with rarefied or cavitated fluids.

The one-dimensional interface is defined by the location of the Lagrangian boundary nodes that are adjacent to grid nodes of the Eulerian mesh. This interface location is used to construct a signed distance function in order to apply level set methods near the interface. Before defining values in the Eulerian ghost nodes, a check is performed to see whether enough ghost nodes are present. That is, since the Lagrangian mesh is moving, one needs to ensure that there is adequate overlap between the two meshes. This is done by examining the values of ϕ on the computational boundaries of the Eulerian mesh. If the computational boundary is an Eulerian ghost node, then the value of ϕ gives the distance to the interface and can be used to estimate the number of ghost nodes that exist between the interface and the computational boundary. Then the size of the Eulerian mesh can be increased if there are not enough ghost nodes.

The Eulerian ghost nodes are defined by first extrapolating S and p. Then u at each ghost node is assigned the value of u at the nearest Lagrangian boundary node that lies on the interface between the Eulerian and Lagrangian grids. Force boundary conditions are applied to the Lagrangian interface using the pressure from the Eulerian grid. First, the pressure at the interface is determined using linear interpolation from the

Eulerian mesh. This linear interpolation requires valid pressure values in both the real and the ghost nodes. Therefore, the pressure extrapolation step needs to be carried out before this linear interpolation step. This Eulerian interface pressure makes a contribution of $\pm p$ to the net force on the Lagrangian boundary node, depending on whether the Lagrangian mesh lies to the right or to the left of the interface, respectively.

With boundary conditions specified on both the Eulerian and Lagrangian meshes, both can be advanced one Euler step in time. Both the Eulerian real grid nodes and a band of Eulerian ghost nodes are advanced in time. These ghost nodes are advanced in time so that they have valid values in case they are uncovered by the Lagrangian mesh.

The two-dimensional interface is defined by the line segments of the Lagrangian mesh boundary that are adjacent to grid nodes of the Eulerian mesh. This interface is used to construct a signed distance function defined at every Eulerian grid node. Once again, ϕ is examined on the computational boundaries of the Eulerian mesh to ensure that enough ghost nodes are present, and the size of the Eulerian mesh is increased when necessary.

The Eulerian ghost nodes are defined by first extrapolating S, p, and $\vec{V}$ across the interface into the ghost nodes. Then for each ghost node, the closest point on the Lagrangian interface is determined. If the closest point happens to be on the end of a linear segment, i.e., a Lagrangian grid node, then that velocity can be designated the closest interface velocity. Otherwise, the closest point is on an edge connecting two Lagrangian grid nodes, and the closest interface velocity is determined using linear interpolation between those two nodes. The normal component of the interface velocity is combined with the tangential component of the extrapolated velocity to determine the velocity at each ghost node. Once the Eulerian ghost nodes have valid values for the extrapolated pressure, force boundary conditions can be determined at the Lagrangian interface. The midpoint of each linear interface segment is defined as a control point, and bilinear interpolation is used to determine the Eulerian mesh pressure at each of these control points. Then this pressure is multiplied by both the length and the inward-pointing normal of the line segment to determine the magnitude and direction of the Eulerian pressure force on this segment. Finally, half of this Eulerian pressure force is added to each of the two nodes that make up the segment.

Figure 17.1 shows the velocity field obtained as a shock wave propagates through an Eulerian gas sandwiched between two Lagrangian materials with strength.

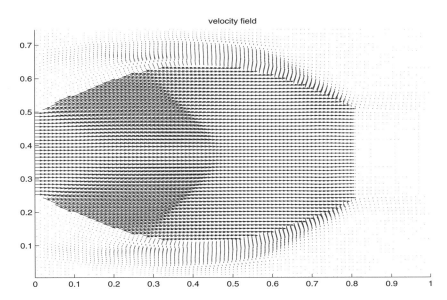

Figure 17.1. A shock wave propagating through a gas bounded on top and bottom by Lagrangian materials with strength. (See also color figure, Plate 10.)

18
Incompressible Flow

18.1 Equations

Starting from conservation of mass, momentum, and energy, the equations for incompressible flow are derived using the divergence-free condition $\nabla \cdot \vec{V} = 0$, which implies that there is no compression or expansion in the flow field. The equation for conservation of mass becomes

$$\rho_t + \vec{V} \cdot \nabla \rho = 0, \tag{18.1}$$

indicating that the (possibly spatially varying) density is advected along streamlines of the flow. The Navier-Stokes equations for viscous incompressible flow are

$$u_t + \vec{V} \cdot \nabla u + \frac{p_x}{\rho} = \frac{(2\mu u_x)_x + (\mu(u_y + v_x))_y + (\mu(u_z + w_x))_z}{\rho}, \tag{18.2}$$

$$v_t + \vec{V} \cdot \nabla v + \frac{p_y}{\rho} = \frac{(\mu(u_y + v_x))_x + (2\mu v_y)_y + (\mu(v_z + w_y))_z}{\rho} + g, \tag{18.3}$$

$$w_t + \vec{V} \cdot \nabla w + \frac{p_z}{\rho} = \frac{(\mu(u_z + w_x))_x + (\mu(v_z + w_y))_y + (2\mu w_z)_z}{\rho}, \tag{18.4}$$

where μ is the viscosity and g is the acceleration of gravity. These equations are more conveniently written in condensed notation as a row vector

$$\vec{V}_t + \left(\vec{V} \cdot \nabla\right) \vec{V} + \frac{\nabla p}{\rho} = \frac{(\nabla \cdot \tau)^T}{\rho} + \vec{g}, \tag{18.5}$$

where "T" is the transpose operator, $\vec{g} = \langle 0, g, 0 \rangle$, and τ is the viscous stress tensor

$$\tau = \mu \begin{pmatrix} 2u_x & u_y + v_x & u_z + w_x \\ u_y + v_x & 2v_y & v_z + w_y \\ u_z + w_x & v_z + w_y & 2w_z \end{pmatrix}, \tag{18.6}$$

which can be expressed in a more compact form as

$$\tau = \mu \begin{pmatrix} \nabla u \\ \nabla v \\ \nabla w \end{pmatrix} + \mu \begin{pmatrix} \nabla u \\ \nabla v \\ \nabla w \end{pmatrix}^T. \tag{18.7}$$

The incompressible flow equations model low-speed fluid events, including many interesting liquid and gas flows observed in everyday life. A number of classic texts have been written about both the analytical and numerical approaches to these equations, including Landau and Lifshitz [101] and Batchelor [11]. A rather inspiring collection of photographs obtained from various experiments was assembled by Van Dyke [169]. A standard introduction to numerical methods for the Navier-Stokes equations is Peyret and Taylor [133], which discusses both the artificial compressibility method of Chorin [46] and the projection method of Chorin [47]. Some of the most popular modern-day numerical methods were introduced by Kim and Moin [96], Bell, Colella, and Glaz [13], and E and Liu [56]. A rather intriguing look at all three of these methods was recently presented by Brown, Cortez, and Minion [19].

18.2   MAC Grid

Harlow and Welch [79] proposed the use of a special grid for incompressible flow computations. This specially defined grid decomposes the computational domain into cells with velocities defined on the cell faces and scalars defined at cell centers. That is, $p_{i,j,k}$, $\rho_{i,j,k}$, and $\mu_{i,j,k}$ are defined at cell centers, while $u_{i \pm \frac{1}{2}, j, k}$, $v_{i, j \pm \frac{1}{2}, k}$, and $w_{i, j, k \pm \frac{1}{2}}$ are defined at the appropriate cell faces.

Equation (18.1) is solved by first defining the cell center velocities with simple averaging:

$$u_{i,j,k} = \frac{u_{i-\frac{1}{2},j,k} + u_{i+\frac{1}{2},j,k}}{2}, \tag{18.8}$$

$$v_{i,j,k} = \frac{v_{i,j-\frac{1}{2},k} + v_{i,j+\frac{1}{2},k}}{2}, \tag{18.9}$$

$$w_{i,j,k} = \frac{w_{i,j,k-\frac{1}{2}} + w_{i,j,k+\frac{1}{2}}}{2}. \tag{18.10}$$

Then the spatial derivatives are evaluated in a straightforward manner, for example using third order accurate Hamilton-Jacobi ENO.

In order to update, u, v, and w on the appropriate cell faces, equations (18.2), (18.3) and (18.4) are written and evaluated on those appropriate cell faces. For example, in order to discretize the convective $\vec{V} \cdot \nabla u$ term at $\vec{x}_{i\pm\frac{1}{2},j,k}$ we first use simple averaging to define $\vec{V}$ at $\vec{x}_{i\pm\frac{1}{2},j,k}$; i.e.,

$$v_{i+\frac{1}{2},j,k} = \frac{v_{i,j-\frac{1}{2},k} + v_{i,j+\frac{1}{2},k} + v_{i+1,j-\frac{1}{2},k} + v_{i+1,j+\frac{1}{2},k}}{4} \tag{18.11}$$

and

$$w_{i+\frac{1}{2},j,k} = \frac{w_{i,j,k-\frac{1}{2}} + w_{i,j,k+\frac{1}{2}} + w_{i+1,j,k-\frac{1}{2}} + w_{i+1,j,k+\frac{1}{2}}}{4} \tag{18.12}$$

define v and w, while u is already defined there. Then the $\vec{V} \cdot \nabla u$ term on the offset $\vec{x}_{i\pm\frac{1}{2},j,k}$ grid is discretized in the same fashion as the $\vec{V} \cdot \nabla \rho$ term on the regular $\vec{x}_{i,j,k}$ grid, for example with third order accurate Hamilton-Jacobi ENO. The convective terms in equations (18.3) and (18.4) are discretized similarly.

The viscous terms are discretized using standard central differencing. For example,

$$(u_x)_{i,j,k} = \frac{u_{i+\frac{1}{2},j,k} - u_{i-\frac{1}{2},j,k}}{\Delta x}, \tag{18.13}$$

$$(u_y)_{i+\frac{1}{2},j+\frac{1}{2},k} = \frac{u_{i+\frac{1}{2},j+1,k} - u_{i+\frac{1}{2},j,k}}{\Delta y}, \tag{18.14}$$

and

$$(u_z)_{i+\frac{1}{2},j,k+\frac{1}{2}} = \frac{u_{i+\frac{1}{2},j,k+1} - u_{i+\frac{1}{2},j,k}}{\Delta z} \tag{18.15}$$

are used to compute the first derivatives of u. The functions ρ and μ are defined only at the grid points, so simple averaging is used to define them elsewhere, e.g.,

$$\mu_{i+\frac{1}{2},j,k} = \frac{\mu_{i,j,k} + \mu_{i+1,j,k}}{2} \tag{18.16}$$

and

$$\mu_{i+\frac{1}{2},j+\frac{1}{2},k} = \frac{\mu_{i,j,k} + \mu_{i+1,j,k} + \mu_{i,j+1,k} + \mu_{i+1,j+1,k}}{4}. \tag{18.17}$$

Then the second-derivative terms are calculated with central differencing as well. For example, $(\mu(u_y + v_x))_y$ in equation (18.2) is discretized as

$$\frac{\mu_{i+\frac{1}{2},j+\frac{1}{2},k}(u_y + v_x)_{i+\frac{1}{2},j+\frac{1}{2},k} - \mu_{i+\frac{1}{2},j-\frac{1}{2},k}(u_y + v_x)_{i+\frac{1}{2},j-\frac{1}{2},k}}{\Delta y} \tag{18.18}$$

at $\vec{x}_{i+\frac{1}{2},j,k}$.

18.3 Projection Method

The projection method is applied by first computing an intermediate velocity field $\vec{V}^\star$, ignoring the pressure term

$$\frac{\vec{V}^\star - \vec{V}^n}{\Delta t} + \left(\vec{V} \cdot \nabla\right)\vec{V} = \frac{(\nabla \cdot \tau)^T}{\rho} + \vec{g}, \qquad (18.19)$$

and then computing a divergence-free velocity field $\vec{V}^{n+1}$,

$$\frac{\vec{V}^{n+1} - \vec{V}^\star}{\Delta t} + \frac{\nabla p}{\rho} = 0, \qquad (18.20)$$

using the pressure as a correction. Note that combining equations (18.19) and (18.20) to eliminate $\vec{V}^\star$ results in equation (18.5) exactly.

Taking the divergence of equation (18.20) results in

$$\nabla \cdot \left(\frac{\nabla p}{\rho}\right) = \frac{\nabla \cdot \vec{V}^\star}{\Delta t} \qquad (18.21)$$

after we set $\nabla \cdot \vec{V}^{n+1}$ to zero, i.e., after we assume that the new velocity field is divergence free. Equation (18.21) defines the pressure in terms of the value of Δt used in equation (18.19). Defining a scaled pressure of $p^\star = p\Delta t$ leads to

$$\vec{V}^{n+1} - \vec{V}^\star + \frac{\nabla p^\star}{\rho} = 0 \qquad (18.22)$$

and

$$\nabla \cdot \left(\frac{\nabla p^\star}{\rho}\right) = \nabla \cdot \vec{V}^\star \qquad (18.23)$$

in place of equations (18.20) and (18.21), where $p^\star$ does not depend on Δt. When the density is spatially constant, we can define $\hat{p} = p\Delta t/\rho$, leading to

$$\vec{V}^{n+1} - \vec{V}^\star + \nabla \hat{p} = 0 \qquad (18.24)$$

and

$$\triangle \hat{p} = \nabla \cdot \vec{V}^\star, \qquad (18.25)$$

where $\hat{p}$ does not depend on Δt or ρ.

Boundary conditions can be applied to either the velocity or the pressure. In order to apply boundary conditions to $\vec{V}^{n+1}$, we apply them to $\vec{V}^\star$ after computing $\vec{V}^\star$ in equation (18.19) and before solving equation (18.21). Then in equation (18.21), we set $\nabla p \cdot \vec{N} = 0$ on the boundary, where $\vec{N}$ is the local unit normal to the boundary. Since the flow is incompressible, the compatibility condition

$$\int_\Gamma \vec{V}^\star \cdot \vec{N} = 0 \qquad (18.26)$$

needs to be satisfied when the boundary condition on $\vec{V}^{\star}$ is specified in order to guarantee the existence of a solution. Here, Γ represents the boundary of the computational domain and $\vec{N}$ is the unit normal to that boundary. See, for example, [133] for more details.

18.4 Poisson Equation

In order to update the intermediate velocity $\vec{V}^{\star}$ obtained with equation (18.19) to the divergence-free $\vec{V}^{n+1}$ using equation (18.20), we need to first find the pressure by solving equation (18.21). This equation is elliptic, since the incompressibility condition is equivalent to assuming an infinite speed of propagation for the sound waves. This elliptic treatment of the acoustic waves gives a CFL condition that depends only on the fluid velocity; i.e., the sound waves do not play a role. Note that one should include the viscosity and gravity terms in the time-step restriction as well, unless, of course, the viscosity is treated implicitly.

The right-hand side of equation (18.21) can be evaluated at each cell center using central differencing; for example,

$$(u_x^{\star})_{i,j,k} = \frac{u_{i+\frac{1}{2},j,k}^{\star} - u_{i-\frac{1}{2},j,k}^{\star}}{\triangle x} \tag{18.27}$$

is used to compute $u_x^{\star}$. The components of the $\nabla p / \rho$ term are computed at cell faces, for example

$$(p_x)_{i+\frac{1}{2},j,k} = \frac{p_{i+1,j,k} - p_{i,j,k}}{\triangle x} \tag{18.28}$$

is used to compute the pressure derivative at $x_{i+\frac{1}{2},j,k}$. This discretization of the pressure derivatives is used both in equation (18.20) to update the velocity field and in equation (18.21) to find the pressure. In equation (18.21) a second derivative is needed as well. Once again, central differencing is used; for example,

$$((p_x/\rho)_x)_{i,j,k} = \frac{(p_x/\rho)_{i+\frac{1}{2},j,k} - (p_x/\rho)_{i-\frac{1}{2},j,k}}{\triangle x} \tag{18.29}$$

is used to compute the second derivative in the x-direction.

Discretization of equation (18.21) at each cell center leads to a coupled linear system of equations for the unknown pressures (one equation at each cell center). This system is symmetric and can be solved in a straightforward manner using a preconditioned conjugate gradient (PCG) method with an incomplete Choleski preconditioner. The interested reader is referred to the comprehensive computational linear algebra text of Golub and Van Loan [75]. After the iterative PCG method is used to compute the pressure at each cell center, the derivatives can be computed at the cell faces and used to update the velocity field in equation (18.20).

18.5   Simulating Smoke for Computer Graphics

For computer graphics applications one is less concerned with overall accuracy and more concerned with computational efficiency, as long as visually believable results can be obtained. For this reason, computer graphics simulations are usually undertaken on relatively coarse grids using methods that allow for large time steps. A popular numerical method in the atmospheric sciences community is the semi-Lagrangian method; see, e.g., Staniforth and Cote [155] for a review. Semi-Lagrangian methods consist in backward tracing of characteristic information from grid points to arbitrary points in space with subsequent interpolation of data from the grid points to the backward-traced origins of the characteristics. This method was first proposed by Courant et al. [51] and is guaranteed to be stable for any time step as long as monotone interpolation is used.

In [70], Fedkiw et al. used a first-order accurate semi-Lagrangian method to produce visually convincing simulations of smoke. Since this method is highly dissipative, the viscous terms were ignored in order to artificially increase the Reynolds number on a coarse grid. To further amplify the numerical Reynolds number, a vorticity confinement term was added as an artificial forcing function on the right-hand side of the equations. This vorticity confinement method was invented by Steinhoff; see, e.g., [161] for more details. Since the air was assumed to have constant density, equation (18.1) was not needed to model the air. On the other hand, since equation (18.1) can be used to model any passive scalar, it was used to independently track the concentration of smoke within the air. Figure 18.1 shows a warm smoke plume injected from left to right rising under the effects of buoyancy, while Figure 18.2 depicts the flow of smoke past a sphere.

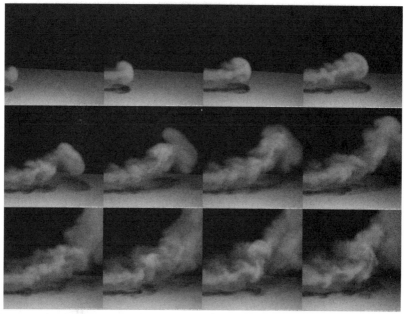

Figure 18.1. A warm smoke plume injected from left to right rises under the influence of buoyancy. (See also color figure, Plate 11.)

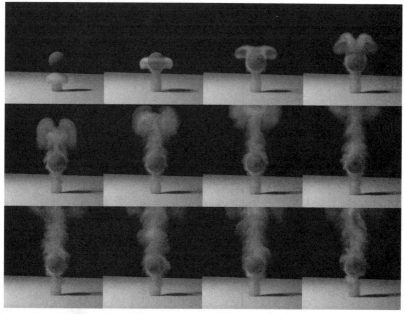

Figure 18.2. Small-scale eddies are generated as smoke flows past a sphere. (See also color figure, Plate 12.)

19
Free Surfaces

19.1 Description of the Model

Consider a two-phase incompressible flow consisting of water and air. The two phase interface separating these fluids has rather complicated dynamics that we will discuss later (in Chapter 21). In many situations the behavior of the heavier fluid significantly dominates the behavior of the lighter fluid. In these situations, the model can be simplified by treating the air as a simple constant-pressure fluid that exerts no other stress (i.e., except pressure forces) on the interface. This removes any relevant dynamic effects from the air, leaving only a simple uniform pressure force normal to the interface independent of the interface topology or motion.

Harlow and Welch [79] used marker particles to identify which grid cells contained water and which grid cells contained air. Later, Raad et al. [136] improved the treatment of the pressure boundary conditions at the interface, introducing a subcell treatment of the pressure. Chen et al. [39] improved the velocity boundary conditions at the free surface to obtain a more accurate flow field. Furthermore, Chen et al. [40] reduced the need to resolve the three-dimensional volume with particles by introducing a method that required particles only near the free surface itself.

Particle locations give only a rough indication of the location of the free surface. We instead prefer to use the level set function ϕ to more accurately model the free surface. Then the Navier-Stokes equations are solved on the water side of the interface only. In order to discretize both the level set equation and the Navier-Stokes equations, ghost cell values of the velocity are needed on the air side of the interface. These are defined by extrapolating the velocity field across the interface using equation (8.1). Dirichlet pressure boundary conditions can be applied at the free surface by setting the pressure to atmospheric pressure at every cell center located in the air. This use of Dirichlet boundary conditions on the pressure reduces the need to enforce the compatibility condition.

The atmospheric pressure boundary condition should be applied directly on the free surface, not at cell centers that are a finite distance away from the surface. Setting the pressure to atmospheric pressure at every cell center in the air causes an overprediction of the pressure at the interface itself. Raad et al. [136] reduces this problem to some degree by using micro cells near the interface, lowering the size of the error constant in this first-order accurate approximation of the boundary condition. Recently, Cheng et al. [44] devised a fully second-order accurate method for enforcing the atmospheric pressure boundary conditions on the free surface. This is accomplished by setting the pressure at cells in the air to specially calculated values that are lower than atmospheric pressure. This implicitly enforces the atmospheric pressure boundary condition exactly on the free surface to second-order accuracy. Notably, this method does not disturb the symmetric nature of the coefficient matrix that needs to be inverted to find the pressures.

19.2 Simulating Water for Computer Graphics

One of the difficulties associated with using level set methods to simulate free surfaces (and fluids in general) is that level set methods tend to suffer from mass loss in underresolved regions of the flow. Foster and Fedkiw [71] addressed this problem in the context of splashing water by devising a new numerical approach that hybridizes the particle method and the level set method using the local interface curvature as a diagnostic. The curvature was used to monitor the interface topology by classifying regions of high curvature as underresolved. In these underresolved regions, the particles are used to rebuild the level set function, reducing mass loss to a large degree. On the other hand, in regions of relatively low curvature the high-order accurate level set solution is preferred. A sample calculation of a splash generated by a sphere is shown in Figure 19.1. In some regions of the flow the grid is too coarse to capture the splashing behavior, even with the aid of the particles, and some particles will inevitably escape from the water side of the interface and appear on the air side. These escaped particles can be used to generate an interesting spray effect, as shown in Figure 19.2, where a slightly submerged ellipse slips through the water.

Although the local interface curvature is a good diagnostic of potential mass loss in level set methods, the approach in [71] is still somewhat ad hoc

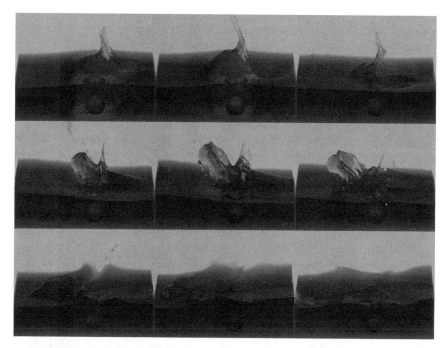

Figure 19.1. A splash is generated as a sphere is thrown into the water. (See also color figure, Plate 13.)

in nature. A more thorough approach to hybridizing particle methods and level set methods (the particle level set method) was presented in Chapter 9. Figure 19.3 shows the highly realistic thin water sheet generated as a sphere splashes into the water. This highlights the ability of the particle level set method to accurately capture thin interfacial regions that may be underresolved by the Cartesian grid. Figure 19.4 shows the impressive results generated using this method to model water flowing into a cylindrical glass. A close-up view of the bottom of the glass is shown in Figure 19.5.

Figure 19.2. An interesting spray effect is generated as a slightly submerged ellipse slips through the water. (See also color figure, Plate 14.)

Figure 19.3. A thin water sheet is generated by a sphere thrown into the water. (See also color figure, Plate 15.)

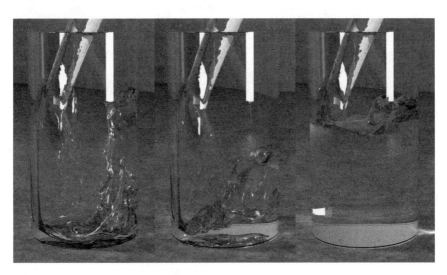

Figure 19.4. Pouring water into a cylindrical glass using the particle level set method. (See also color figure, Plate 16.)

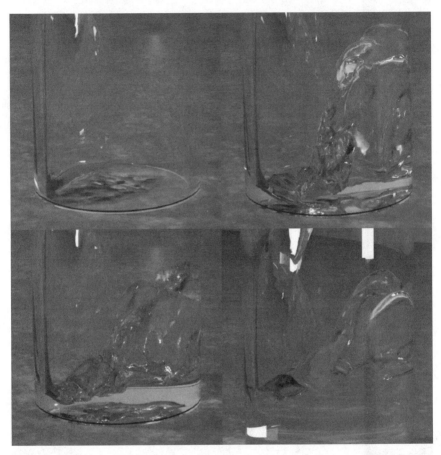

Figure 19.5. Pouring water into a cylindrical glass using the particle level set method. (See also color figure, Plate 17.)

20
Liquid-Gas Interactions

20.1 Modeling

Liquid-gas interactions, e.g., the vaporization and subsequent combustion of liquid fuel droplets or the shock-induced mixing of liquids, are rather difficult problems in computational fluid dynamics. These problems address the interaction of liquid droplets with a compressible gas medium. There are three classical approaches to such problems: Both phases can be treated as compressible fluids (as we did in Chapter 15), both phases can be treated as incompressible fluids (as we did in Chapter 21), or the gas can be treated as a compressible fluid while the liquid is treated as an incompressible fluid. A completely incompressible treatment can be ruled out any time one is interested in shock waves or other compressible phenomena. A completely compressible treatment is not desirable, since a relatively high sound speed in the liquid phase can impose a restrictive (and inefficient) CFL condition. Moreover, a completely compressible approach is limited to liquids for which there are acceptable models for their compressible evolution. To overcome these difficulties, Caiden et al. [23] modeled the gas as a compressible fluid and the liquid as an incompressible fluid. They coupled a high-resolution shock-capturing scheme for the compressible gas flow to a standard incompressible flow solver for the liquid phase.

20.2   Treating the Interface

Since the interface is a contact discontinuity moving with the local fluid velocity, the Rankine-Hugoniot jump conditions imply that both the pressure and the normal velocity are continuous across the interface. An incompressible liquid can be thought of as the limiting case obtained by increasing the sound speed of a compressible liquid to infinity. In this sense, an incompressible fluid can be thought of as a very stiff compressible fluid, in fact, the stiffest. The interface separating the compressible flow from the incompressible flow is treated using the robust interpolation procedure outlined in Section 15.9. That is, the compressible gas is used to determine the interface pressure, while the incompressible liquid is used to determine the interface normal velocity.

Advancing the solution forward in time consists of four steps. First, the entire incompressible velocity field is extrapolated across the interface. The ghost cell values are used to find the intermediate incompressible velocity field $\vec{V}^\star$. Second, the entire compressible state vector is extrapolated across the interface, and the extrapolated tangential velocity is combined with the incompressible normal velocity to obtain a ghost cell velocity for the compressible fluid. Then the compressible gas is updated in time. Third, the level set function is advanced forward in time using the incompressible velocity field only, since the interface velocity is defined by the incompressible flow. The extrapolated ghost cell values of the incompressible velocity field are useful in this step. Fourth, the intermediate incompressible flow velocity $\vec{V}^\star$ is projected into a divergence-free state using the updated level set location and the updated values of the compressible pressure as Dirichlet boundary conditions at the interface. This last step accounts for the interface forces imposed by the pressure of the compressible fluid. Surface tension effects are easily included in this last step using Dirichlet pressure boundary conditions of $p = p_c + \sigma\kappa$, where p_c is the compressible pressure, σ is a constant, and κ is the local interface curvature. This accounts for the jump in pressure due to surface tension forces, i.e., $[p] = \sigma\kappa$.

Figure 20.1 shows an incompressible liquid droplet moving from left to right in a compressible gas flow. Notice the lead shock in the compressible gas. Figure 20.2 shows a shock wave impinging on an incompressible liquid droplet. A reflected wave can be seen to the left, and a faint transmitted wave can be seen to the right.

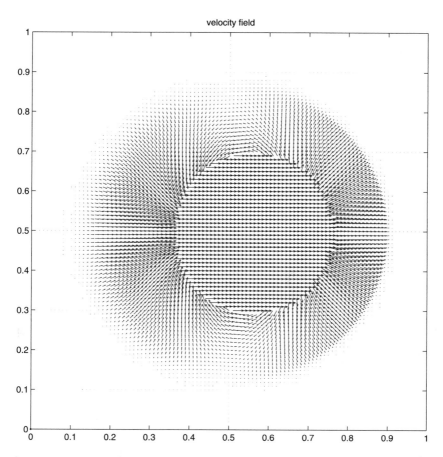

Figure 20.1. An incompressible droplet traveling to the right in a compressible gas flow. Note the lead shock wave. (See also color figure, Plate 18.)

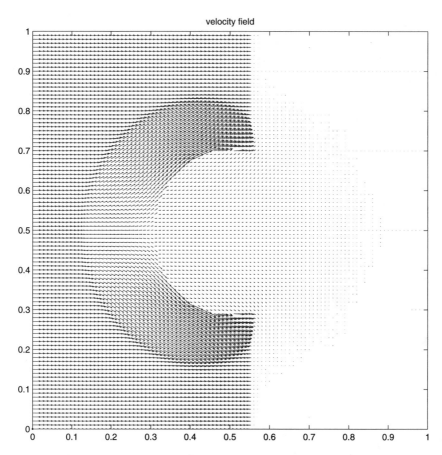

Figure 20.2. A shock wave impinging on an incompressible droplet producing a reflected wave and a (very) weak transmitted wave. (See also color figure, Plate 19.)

21
Two-Phase Incompressible Flow

21.1 Introduction

The earliest real success in the coupling of the level set method to problems involving external physics came in computing two-phase incompressible flow, in particular see the work of Sussman et al. [160] and Chang et al. [38]. In two-phase incompressible flow, the Navier-Stokes equations are used to model the fluids on both sides of the interface. Generally, the fluids have different densities and viscosities, so these quantities are discontinuous across the interface. Both the discontinuous viscosity and surface tension forces cause the pressure to be discontinuous across the interface as well. In addition, a discontinuous viscosity leads to kinks in the velocity field across the interface.

In [132], Peskin introduced the "immersed boundary" method for simulating an elastic membrane immersed in an incompressible fluid flow. This method uses a δ-function formulation to smear out the numerical solution in a thin region about the immersed interface. This concept has been used by a variety of authors to solve a number of interface-related problems. For example, [160] defined a numerically smeared-out density and viscosity as functions of the level set function,

$$\rho(\phi) = \rho^- + \left(\rho^+ - \rho^-\right) H(\phi), \tag{21.1}$$

$$\mu(\phi) = \mu^- + \left(\mu^+ - \mu^-\right) H(\phi), \tag{21.2}$$

where $H(\phi)$ is the numerically smeared-out Heaviside function defined by equation (1.22). This removes all discontinuities across the interface, except the jump in pressure due to surface tension, $[p] = \sigma\kappa$, where σ is a constant coefficient and κ is the local curvature of the interface. Using the immersed boundary method to smear out the pressure across the interface leads to continuity of the pressure, $[p] = 0$, and loss of all surface-tension effects. This was remedied by Brackbill et al. [18] in the context of volume of fluid (VOF) methods and by Unverdi and Tryggvason [168] in the context of front-tracking methods by adding a new forcing term to the right-hand side of the momentum equations. In the context of level set methods (see [160]) this new forcing term takes the form

$$\frac{\delta(\phi)\sigma\kappa\vec{N}}{\rho}, \tag{21.3}$$

where $\delta(\phi)$ is the smeared-out delta function given by equation (1.23). In the spirit of the immersed boundary method, [18] referred to this as the continuum surface force (CSF) method.

In the interest of solving for the pressure jump directly, Liu et al. [106] devised a new boundary-condition-capturing approach for the variable-coefficient Poisson equation to solve problems of the form

$$\nabla\left(\frac{1}{\rho}\nabla p\right) = f, \tag{21.4}$$

where the jump conditions $[p] = g$ and $[(1/\rho)\nabla p \cdot \vec{N}] = h$ are given. Here, ρ can be discontinuous across the interface as well. Figure 21.1 shows a typical example of the discontinuous solutions obtained using this method. Note that both the pressure and its derivatives are clearly discontinuous across the interface. Kang et al. [91] applied this technique to multiphase incompressible flow, illustrating the ability to solve these equations without smearing out the density, the viscosity, or the pressure across the interface. Moreover, the $\delta(\phi)\sigma\kappa\vec{N}/\rho$ forcing term was not needed, since the pressure jump was modeled directly. Figure 21.2 shows a water drop falling through the air into the water. Here, surface tension forces cause the spherically shaped region at the top of the resulting water jet in the last frame of the figure.

LeVeque and Li [102] proposed a second-order accurate sharp interface method to solve equation (21.4). In general, one needs to solve a linear system of equations of the form $A\vec{p} = \vec{b}$, where $\vec{p}$ are the unknown pressures, A is the coefficient matrix, and $\vec{b}$ is the right-hand side. Unfortunately, the discretization in [102] leads to a complicated asymmetric coefficient matrix, making this linear system difficult to solve. So far, this method has not been applied to two-phase incompressible flow. In contrast, the discretization proposed in [106] leads to a symmetric coefficient matrix identical to the standard one obtained when both the pressure and its derivatives are continuous across the interface. Adding the jump conditions only requires modification of the right-hand side, $\vec{b}$. This allows the use of standard

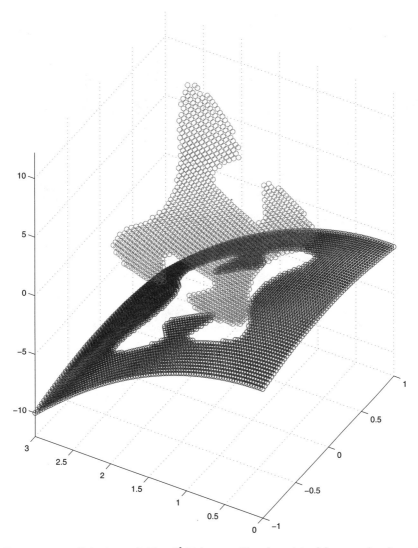

Figure 21.1. Solution of $\nabla \cdot \left(\frac{1}{\rho}\nabla p\right) = f(x,y)$, with $[p] = g(x,y)$ and $\left[\frac{1}{\rho}\nabla p \cdot \vec{N}\right] = h(x,y)$. Note the sharp resolution of the discontinuity.

black-box linear system solvers even in the presence of complicated jump conditions. Recently, Li and Lai [103] extended [106] by adding a second-order accurate correction term to the method. This correction term is valid in the presence of an immersed interface, raising the order of accuracy from one to two in that instance. Unfortunately, a correction term for two-phase incompressible flow does not yet exist.

Figure 21.2. A water drop falls through the air into the water. Surface tension forces cause the spherically shaped region at the top of the water jet in the last frame. (See also color figure, Plate 20.)

21.2 Jump Conditions

Applying conservation to the two-phase incompressible flow interface results in the jump conditions

$$\left[\begin{pmatrix} \vec{N} \\ \vec{T_1} \\ \vec{T_2} \end{pmatrix} (pI - \tau)\vec{N}^T \right] = \begin{pmatrix} \sigma \kappa \\ 0 \\ 0 \end{pmatrix}, \tag{21.5}$$

where $\vec{T_1}$ and $\vec{T_2}$ are orthogonal unit tangent vectors, I is the identity matrix, and τ is the viscous stress tensor; see equation (18.6). Equation (21.5) states that the net stress on the interface must be zero, since it has no mass.

Since the flow is viscous, the velocities and their tangential derivatives are continuous across the interface, i.e.,

$$[u] = [v] = [w] = 0, \tag{21.6}$$

$$[\nabla u \cdot \vec{T_1}] = [\nabla v \cdot \vec{T_1}] = [\nabla w \cdot \vec{T_1}] = 0, \tag{21.7}$$

$$[\nabla u \cdot \vec{T_2}] = [\nabla v \cdot \vec{T_2}] = [\nabla w \cdot \vec{T_2}] = 0. \tag{21.8}$$

This leads to the jump condition

$$\left[\left(\nabla u \cdot \vec{N}, \nabla v \cdot \vec{N}, \nabla w \cdot \vec{N}\right) \cdot \vec{N}\right] = 0, \tag{21.9}$$

which states that the normal derivative of the normal component of the velocity field is continuous across the interface. Using this, the jump condition

$$[p] - 2\,[\mu]\left(\nabla u \cdot \vec{N}, \nabla v \cdot \vec{N}, \nabla w \cdot \vec{N}\right) \cdot \vec{N} = \sigma\kappa \tag{21.10}$$

can be written for the pressure. Notice that this reduces to $[p] = \sigma\kappa$ when the viscosity is continuous across the interface. Further derivations lead to

$$
\begin{aligned}
&\begin{pmatrix} [\mu u_x] & [\mu u_y] & [\mu u_z] \\ [\mu v_x] & [\mu v_y] & [\mu v_z] \\ [\mu w_x] & [\mu w_y] & [\mu w_z] \end{pmatrix} \\
&= [\mu]\left(\begin{pmatrix} \nabla u \\ \nabla v \\ \nabla w \end{pmatrix} \begin{pmatrix} \vec{0} \\ \vec{T_1} \\ \vec{T_2} \end{pmatrix}^T \begin{pmatrix} \vec{0} \\ \vec{T_1} \\ \vec{T_2} \end{pmatrix}\right. \\
&\qquad + \vec{N}^T\vec{N} \begin{pmatrix} \nabla u \\ \nabla v \\ \nabla w \end{pmatrix} \vec{N}^T\vec{N} \\
&\qquad \left. - \begin{pmatrix} \vec{0} \\ \vec{T_1} \\ \vec{T_2} \end{pmatrix}^T \begin{pmatrix} \vec{0} \\ \vec{T_1} \\ \vec{T_2} \end{pmatrix} \begin{pmatrix} \nabla u \\ \nabla v \\ \nabla w \end{pmatrix}^T \vec{N}^T\vec{N}\right),
\end{aligned}
\tag{21.11}
$$

which is useful for discretizing the viscous terms, especially since the right-hand side of this equation involves only derivatives that are continuous across the interface. Notice that all the quantities on the left-hand side of this equation become continuous across the interface when $[\mu] = 0$ forces the right-hand side to be identically zero. See [106] for details.

Since the velocity is continuous across the interface, the material derivative, or Lagrangian, acceleration is continuous as well,

$$\left[\frac{Du}{Dt}\right] = \left[\frac{Dv}{Dt}\right] = \left[\frac{Dw}{Dt}\right] = 0. \tag{21.12}$$

Since the Navier-Stokes equations (18.2), (18.3), and (18.4) are valid on both sides of the interface, these equations do not jump across the interface; i.e.,

$$\left[\vec{V_t} + \left(\vec{V} \cdot \nabla\right)\vec{V} + \frac{\nabla p}{\rho} - \frac{(\nabla \cdot \tau)^T}{\rho} - \vec{g}\right] = 0. \tag{21.13}$$

This can be combined with equation (21.12) to obtain

$$\left[\frac{\nabla p}{\rho}\right] = \left[\frac{(\nabla \cdot \tau)^T}{\rho}\right],$$

(21.14)

which is equivalent to

$$\left[\frac{p_x}{\rho}\right] = \left[\frac{(2\mu u_x)_x + (\mu(u_y + v_x))_y + (\mu(u_z + w_x))_z}{\rho}\right],$$

(21.15)

$$\left[\frac{p_y}{\rho}\right] = \left[\frac{(\mu(u_y + v_x))_x + (2\mu v_y)_y + (\mu(v_z + w_y))_z}{\rho}\right],$$

(21.16)

$$\left[\frac{p_z}{\rho}\right] = \left[\frac{(\mu(u_z + w_x))_x + (\mu(v_z + w_y))_y + (2\mu w_z)_z}{\rho}\right],$$

(21.17)

in expanded form.

The two-phase incompressible flow equations are discretized in the same manner as the equations for one-phase incompressible flow. First, an intermediate velocity field $\vec{V}^{\star}$ is computed using equation (18.19). Then equation (18.21) is solved to find the pressure, which is used in equation (18.20) to make the velocity field divergence free. Due to the discontinuous nature of several quantities across the interface, special care is needed in discretizing the viscous terms in equation (18.19) and in discretizing the pressure and density in equations (18.21) and (18.20).

21.3   Viscous Terms

In the δ-function approach the density and the viscosity are numerically smeared out, so that they are continuous across the interface. The continuous viscosity simplifies the jump conditions, allowing the viscous terms to be discretized just as they were for one-phase incompressible flow. The only difference is that ρ and μ are defined by equations (21.1) and (21.2) respectively. Averaging of ϕ is preferred to averaging of other quantities. For example, the viscosity at $\vec{x}_{i+\frac{1}{2},j,k}$ is defined as

$$\mu_{i+\frac{1}{2},j,k} = \mu\left(\frac{\phi_{i,j,k} + \phi_{i+1,j,k}}{2}\right),$$

(21.18)

as opposed to

$$\mu_{i+\frac{1}{2},j,k} = \frac{\mu(\phi_{i,j,k}) + \mu(\phi_{i+1,j,k})}{2},$$

(21.19)

as was done in equation (18.16).

If one prefers to keep the density and viscosity sharp across the interface, the sign of the level set function can be used to determine μ as μ^- or μ^+

and to determine ρ as ρ^- or ρ^+. Consider the case where μ and ρ are both independently spatially constant on either side of the interface, allowing the simplification of the viscous terms to

$$\frac{\mu \left(u_{xx} + u_{yy} + u_{zz} \right)}{\rho}, \tag{21.20}$$

$$\frac{\mu \left(v_{xx} + v_{yy} + v_{zz} \right)}{\rho}, \tag{21.21}$$

$$\frac{\mu \left(w_{xx} + w_{yy} + w_{zz} \right)}{\rho}, \tag{21.22}$$

with the aid of $\nabla \cdot \vec{V} = 0$. Since the velocities are continuous, their first derivatives can be computed directly. However, the jump conditions in equation (21.11) are needed to compute the second derivatives.

The right-hand side of equation (21.11) needs to be computed in order to evaluate the jumps across the interface. First, the continuous velocity field is averaged from the MAC grid to the cell centers. Then central differencing is used to compute the first derivatives at each cell center. These first derivatives are multiplied by the appropriate components of the normal and tangent vectors to obtain a numerical estimate for the right-hand side of equation (21.11), which we denote by the matrix J. Since J is a spatially continuous function, spatial averages can be used to define J elsewhere. For example, $J_{i+\frac{1}{2},j,k} = \left(J_{i,j,k} + J_{i+1,j,k} \right) / 2$.

Once J has been computed, the second derivatives are computed using techniques similar to those developed in [106]. For example, consider the discretization of μu_{xx} at $x_{i+\frac{1}{2},j,k}$ using $u_M = u_{i+\frac{1}{2},j,k}$ and its neighbors $u_L = u_{i-\frac{1}{2},j,k}$ and $u_R = u_{i+\frac{3}{2},j,k}$. We need averaged values of ϕ and J^{11} (the appropriate scalar entry of J) at the same three spatial locations as the u terms. If ϕ_L, ϕ_M, and ϕ_R are all greater than zero, we define

$$(\mu u_x)_L = \mu^+ \left(\frac{u_M - u_L}{\Delta x} \right) \tag{21.23}$$

and

$$(\mu u_x)_R = \mu^+ \left(\frac{u_R - u_M}{\Delta x} \right), \tag{21.24}$$

arriving at

$$(\mu u_{xx})_{i+\frac{1}{2},j,k} = \frac{(\mu u_x)_R - (\mu u_x)_L}{\Delta x} \tag{21.25}$$

in the standard fashion. A similar discretization holds when all three ϕ values are less than or equal to zero.

Suppose that $\phi_L \leq 0$ and $\phi_M > 0$, so that the interface lies between the associated grid points. Then

$$\theta = \frac{|\phi_L|}{|\phi_L| + |\phi_M|} \tag{21.26}$$

can be used to estimate the interface location by splitting this cell into two pieces of size $\theta \triangle x$ on the left and $(1 - \theta)\triangle x$ on the right. At the interface, we denote the continuous velocity by u_I and calculate the jump as $J_I = \theta J_M + (1-\theta)J_L$. Then we discretize the jump condition $[\mu u_x] = J_I$ as

$$\mu^+ \left(\frac{u_M - u_I}{(1 - \theta)\triangle x} \right) - \mu^- \left(\frac{u_I - u_L}{\theta \triangle x} \right) = J_I, \tag{21.27}$$

solving for u_I to obtain

$$u_I = \frac{\mu^+ u_M \theta + \mu^- u_L(1 - \theta) - J_I \theta(1 - \theta)\triangle x}{\mu^+ \theta + \mu^-(1 - \theta)}, \tag{21.28}$$

so that we can write

$$(\mu u_x)_L = \mu^+ \left(\frac{u_M - u_I}{(1 - \theta)\triangle x} \right) = \hat{\mu} \left(\frac{u_M - u_L}{\triangle x} \right) + \frac{\hat{\mu} J_I \theta}{\mu^-}, \tag{21.29}$$

where

$$\hat{\mu} = \frac{\mu^+ \mu^-}{\mu^+ \theta + \mu^-(1 - \theta)} \tag{21.30}$$

defines an effective μ. Similarly, if $\phi_L > 0$ and $\phi_M \leq 0$, then

$$(\mu u_x)_L = \mu^- \left(\frac{u_M - u_I}{(1 - \theta)\triangle x} \right) = \hat{\mu} \left(\frac{u_M - u_L}{\triangle x} \right) - \frac{\hat{\mu} J_I \theta}{\mu^+}, \tag{21.31}$$

where

$$\hat{\mu} = \frac{\mu^- \mu^+}{\mu^- \theta + \mu^+(1 - \theta)} \tag{21.32}$$

defines an effective μ.

In similar fashion, if $\phi_R > 0$ and $\phi_M \leq 0$, then

$$\theta = \frac{|\phi_R|}{|\phi_R| + |\phi_M|} \tag{21.33}$$

is used to estimate the interface location with $(1 - \theta)\triangle x$ on the left and $\theta \triangle x$ on the right. Then $J_I = \theta J_M + (1 - \theta)J_R$ is used to discretize the jump condition as

$$\mu^+ \left(\frac{u_R - u_I}{\theta \triangle x} \right) - \mu^- \left(\frac{u_I - u_M}{(1 - \theta)\triangle x} \right) = J_I, \tag{21.34}$$

resulting in

$$u_I = \frac{\mu^- u_M \theta + \mu^+ u_R(1 - \theta) - J_I \theta(1 - \theta)\triangle x}{\mu^- \theta + \mu^+(1 - \theta)} \tag{21.35}$$

and

$$(\mu u_x)_R = \mu^- \left(\frac{u_I - u_M}{(1-\theta)\Delta x} \right) = \hat{\mu} \left(\frac{u_R - u_M}{\Delta x} \right) - \frac{\hat{\mu} J_I \theta}{\mu^+}, \qquad (21.36)$$

where

$$\hat{\mu} = \frac{\mu^- \mu^+}{\mu^- \theta + \mu^+ (1-\theta)} \qquad (21.37)$$

defines an effective μ. If $\phi_R \leq 0$ and $\phi_M > 0$, then

$$(\mu u_x)_R = \mu^+ \left(\frac{u_I - u_M}{(1-\theta)\Delta x} \right) = \hat{\mu} \left(\frac{u_R - u_M}{\Delta x} \right) + \frac{\hat{\mu} J_I \theta}{\mu^-}, \qquad (21.38)$$

where

$$\hat{\mu} = \frac{\mu^+ \mu^-}{\mu^+ \theta + \mu^- (1-\theta)} \qquad (21.39)$$

defines an effective μ.

21.4 Poisson Equation

Consider solving

$$\nabla \cdot (\beta \nabla p) = f \qquad (21.40)$$

for the pressure p with specified jump conditions of $[p] = a$ and $[\beta p_n] = b$ across the interface. In the context of equation (18.21), $\beta = 1/\rho$ and $f = (\nabla \cdot \vec{V}^\star)/\Delta t$. When the δ-function method is used, this equation is straightforward to solve, since a and b are set to 0, and equation (21.1) is used to define a smeared-out continuous value for ρ. In this case, one obtains a numerically smeared-out pressure profile that does not include surface-tension forces. As discussed above, the forcing function defined in equation (21.3) can be added to the momentum equations to recover surface-tension effects. On the other hand, if one wants to model the surface-tension effects directly, the jump conditions for the pressure cannot be ignored.

First, consider the one-dimensional problem where a standard second-order accurate discretization of

$$\frac{\beta_{i+\frac{1}{2}} \left(\frac{p_{i+1} - p_i}{\Delta x} \right) - \beta_{i-\frac{1}{2}} \left(\frac{p_i - p_{i-1}}{\Delta x} \right)}{\Delta x} = f_i \qquad (21.41)$$

can be written for each unknown p_i. Suppose that the interface is located between x_k and x_{k+1}. As in the treatment of the viscosity term, we discretize the jump condition $[\beta p_x] = b$, obtaining, for example,

$$\beta^+ \left(\frac{p_{k+1} - p_I}{(1-\theta)\Delta x} \right) - \beta^- \left(\frac{p_I - p_k}{\theta \Delta x} \right) = b, \qquad (21.42)$$

and solve for p_I as

$$p_I = \frac{\beta^+ p_{k+1}\theta + \beta^- p_k(1-\theta) - b\theta(1-\theta)\Delta x}{\beta^+\theta + \beta^-(1-\theta)}, \quad (21.43)$$

so that approximations to the derivatives on the left and right sides of the interface can be written as

$$\beta^-\left(\frac{p_I - p_k}{\theta\Delta x}\right) = \hat{\beta}\left(\frac{p_{k+1} - p_k}{\Delta x}\right) - \frac{\hat{\beta}b(1-\theta)}{\beta^+} \quad (21.44)$$

and

$$\beta^+\left(\frac{p_{k+1} - p_I}{(1-\theta)\Delta x}\right) = \hat{\beta}\left(\frac{p_{k+1} - p_k}{\Delta x}\right) + \frac{\hat{\beta}b\theta}{\beta^-}, \quad (21.45)$$

where

$$\hat{\beta} = \frac{\beta^+\beta^-}{\beta^+\theta + \beta^-(1-\theta)} \quad (21.46)$$

defines an effective β.

The new equations for the unknowns p_k and p_{k+1} are then

$$\frac{\hat{\beta}\left(\frac{(p_{k+1}-a)-p_k}{\Delta x} - \frac{b(1-\theta)}{\beta^+}\right) - \beta_{k-\frac{1}{2}}\left(\frac{p_k - p_{k-1}}{\Delta x}\right)}{\Delta x} = f_k \quad (21.47)$$

and

$$\frac{\beta_{k+\frac{3}{2}}\left(\frac{p_{k+2}-p_{k+1}}{\Delta x}\right) - \hat{\beta}\left(\frac{u_{k+1}-(p_k+a)}{\Delta x} + \frac{b\theta}{\beta^-}\right)}{\Delta x} = f_{k+1}, \quad (21.48)$$

where we add the a term to correct for the fact the pressure is discontinuous across the interface as well. Note that this correction was not necessary in treating the viscosity, since the velocity field is continuous across the interface. These new equations for p_k and p_{k+1} can be rewritten in standard form as

$$\frac{\hat{\beta}\left(\frac{p_{k+1}-p_k}{\Delta x}\right) - \beta_{k-\frac{1}{2}}\left(\frac{p_k - p_{k-1}}{\Delta x}\right)}{\Delta x} = f_k + \frac{\hat{\beta}a}{(\Delta x)^2} + \frac{\hat{\beta}b(1-\theta)}{\beta^+\Delta x} \quad (21.49)$$

and

$$\frac{\beta_{k+\frac{3}{2}}\left(\frac{p_{k+2}-p_{k+1}}{\Delta x}\right) - \hat{\beta}\left(\frac{p_{k+1}-p_k}{\Delta x}\right)}{\Delta x} = f_{k+1} - \frac{\hat{\beta}a}{(\Delta x)^2} + \frac{\hat{\beta}b\theta}{\beta^-\Delta x} \quad (21.50)$$

emphasizing that this discretization yields the standard symmetric linear system with $\beta_{k+1/2} = \hat{\beta}$.

More generally, at each grid point i, we write a linear equation of the form

$$\frac{\beta_{i+\frac{1}{2}}\left(\frac{p_{i+1}-p_i}{\Delta x}\right) - \beta_{i-\frac{1}{2}}\left(\frac{p_i-p_{i-1}}{\Delta x}\right)}{\Delta x} = f_i + F^L + F^R \quad (21.51)$$

and assemble the system of linear equations into matrix form. Each $\beta_{k+1/2}$ is evaluated based on the side of the interface that x_k and x_{k+1} lie on, and a special $\hat{\beta}$ is used when x_k and x_{k+1} lie on opposite sides of the interface. Then if the left arm of the stencil (the line segment connecting x_i and x_{i-1}) crosses the interface, a nonzero F^L is defined with correction terms for $[p] = a$ and $[\beta p_n] = b$. Likewise, if the right arm of the stencil (the line segment connecting x_i and x_{i+1}) crosses the interface, a nonzero F^R is defined with correction terms for $[p] = a$ and $[\beta p_n] = b$.

The multidimensional approach is treated in a dimension-by-dimension fashion. While the $[p] = a$ jump condition is trivial to apply, some assumptions are made in order to obtain a dimension-by-dimension approach for $[\beta p_n] = b$. For example, in two spatial dimensions we assume that

$$[\beta p_x] = [\beta p_n]n_1 \tag{21.52}$$

and

$$[\beta p_y] = [\beta p_n]n_2, \tag{21.53}$$

where n_1 and n_2 are the components of the local unit normal. Although these equations are not generally true, adding n_1 times the first equation to n_2 times the second equation leads to the correct $[\beta p_n] = b$ jump condition and numerically demonstrated convergence to the correct solution. The errors in this approach can be characterized by adding t_1 times the first equation to t_2 times the second equation to obtain $[\beta p_t] = 0$, implying that the tangential derivative is incorrectly smeared out. The two-dimensional application of the method consists in writing a linear equation of the form

$$\frac{\beta_{i+1/2,j}\left(\frac{p_{i+1,j}-p_{i,j}}{\Delta x}\right) - \beta_{i-1/2,j}\left(\frac{p_{i,j}-p_{i-1,j}}{\Delta x}\right)}{\Delta x}$$

$$+ \frac{\beta_{i,j+1/2}\left(\frac{p_{i,j+1}-p_{i,j}}{\Delta y}\right) - \beta_{i,j-1/2}\left(\frac{p_{i,j}-p_{i,j-1}}{\Delta y}\right)}{\Delta y}$$

$$= f_{i,j} + F^x + F^y \tag{21.54}$$

at each grid point, where $F^x = F^L + F^R$ and $F^y = F^B + F^T$ are obtained by considering each spatial dimension independently using either $[\beta p_x] = [\beta p_n]n_1 = bn_1$ or $[\beta p_y] = [\beta p_n]n_2 = bn_2$, respectively.

Before using the above-described numerical method to solve equation (18.21), the jump condition given in equation (21.10) needs to be computed. This can be done with standard central differencing of the averaged cell-centered velocities, analogous to the way that J was computed in discretizing the viscous terms. Note that we can set $[p_x/\rho] = [p_y/\rho] = [p_z/\rho] = 0$ in spite of the nonzero jumps in these quantities. Since the full equations are continuous across the interface, one can take the divergence of the full equations to derive equation (18.21) without the need for correction terms. The jumps in the derivatives of the pressure in equation (18.21) are already balanced out on the right-hand side by the appropriate jumps included in the V^* term.

The resulting linear system of equations can still be solved using a PCG gradient with an incomplete Choleski preconditioner, just as in the case of one-phase incompressible flow. However, one needs to use caution when plugging the resulting pressure into equation (18.20), since the pressure is discontinuous across the interface. The pressure derivatives in equation (18.20) should be computed in exactly the same fashion as they were computed in solving equation (18.21); i.e., the correction terms are still needed.

22
Low-Speed Flames

22.1   Reacting Interfaces

In Chapter 21 the interface moved with the local fluid velocity only, and individual fluid particles did not cross the interface. In this chapter we consider interfaces across which a chemical reaction is converting one incompressible fluid into another. The interface moves with the local velocity of the unreacted fluid plus a reaction term that accounts for the conversion of one fluid into the other as material moves across the interface. Consider an interface separating liquid and gas regions where the liquid is actively vaporizing into the gaseous state. Juric and Tryg-gvason [90] developed a front-tracking approach to this problem using a δ-function formulation to treat the interface boundary conditions. Son and Dir [153] and Welch and Wilson [171] developed level-set-based and volume-of-fluid-based (respectively) approaches to this same problem also using a δ-function formulation.

Another example of reacting interfaces occurs in premixed flames. Assuming an infinitely thin flame front allows us to treat the flame front as a discontinuity separating two incompressible flows. The unreacted material undergoes reaction as it crosses the interface, producing a lower-density (higher-volume) reacted material. Qian et al. [135] devised a front-tracking approach to this problem using a δ-function formulation.

Typically, the density is discontinuous across the interface. Thus, material must instantaneously expand as it crosses the interface, implying that the normal velocity is discontinuous across the interface as well (in addition to the discontinuity of the density, viscosity, and pressure). Delta-function formulations smear out this velocity jump, forcing a continuous velocity field across the interface. This can be problematic, since this numerical smearing adds a compressible character to the flow field near the interface. The divergence-free condition is not exactly satisfied in the separate subdomains. In addition, difficulties arise in computing the interface velocity, which depends on the local velocity of the unreacted material. Near the interface, the velocity of the unreacted material contains large $O(1)$ numerical errors where it has been nonphysically forced to be continuous with the velocity of the reacted material. Partial solutions to these problems were proposed by Helenbrook et al. [84], where the authors were able to remove the numerical smearing of the normal velocity, obtaining a sharp interface profile. This method works well as long as flame fronts remain well separated with moderate curvature; see Helenbrook and Law [83]. This method cannot treat merging flame fronts or individual fronts with relatively high curvature. These drawbacks were recently overcome by Nguyen et al. [119] who extended the work of Kang et al. [91] to treat this problem.

22.2   Governing Equations

We ignore viscous effects and consider the equations for inviscid incompressible flow

$$\vec{V}_t + \left(\vec{V} \cdot \nabla\right)\vec{V} + \frac{\nabla p}{\rho} = 0 \tag{22.1}$$

independently for each fluid. The interface velocity is $\vec{W} = D\vec{N}$, where D is the normal component of the interface velocity defined by $D = (V_N)_u + S$. The "u" subscript indicates that the normal velocity is calculated using the velocity of the unreacted material only. This is important to note, since V_N is discontinuous across the interface. The flame speed is defined as $S = S_o + \sigma\kappa$ where, S_o and σ are constants and κ is the local curvature of the interface.

Conservation of mass and momentum imply the standard Rankine-Hugoniot jump conditions across the interface

$$[\rho(V_N - D)] = 0 \tag{22.2}$$
$$[\rho(V_N - D)^2 + p] = 0 \tag{22.3}$$

as well as continuity of the tangential velocities, $[V_{T_1}] = [V_{T_2}] = 0$, as long as $S \neq 0$. Note that $S = 0$ only in the case of a contact discontinuity (not a flame). Denoting the mass flux in the moving reference frame (speed D) by

$$M = \rho_r\left((V_N)_r - D\right) = \rho_u\left((V_N)_u - D\right) \tag{22.4}$$

allows us to rewrite equation (22.2) as $[M] = 0$. Here the "r" subscript denotes a reacted material quantity. Substitution of $D = (V_N)_u + S$ into equation (22.4) yields

$$M = -\rho_u S, \tag{22.5}$$

which is a rather simple quantity for computations.

Starting with $[D] = 0$, we derive

$$\left[\frac{\rho V_N - \rho(V_N - D)}{\rho}\right] = 0, \tag{22.6}$$

$$\left[\frac{\rho V_N - M}{\rho}\right] = 0, \tag{22.7}$$

and

$$[V_N] = M\left[\frac{1}{\rho}\right], \tag{22.8}$$

where the last equation follows from $[M] = 0$. It is more convenient to write

$$[\vec{V}] = M\left[\frac{1}{\rho}\right]\vec{N} \tag{22.9}$$

as a summary of equation (22.8) and $[V_{T_1}] = [V_{T_2}] = 0$. The dot product of equation (22.9) and $\vec{N}$ results in equation (22.8), while the dot product of equation (22.9) and $\vec{T}_1$ or $\vec{T}_2$ results in $[V_{T_1}] = 0$ or $[V_{T_2}] = 0$, respectively. Equation (22.3) can be rewritten as

$$\left[\frac{M^2}{\rho} + p\right] = 0 \tag{22.10}$$

or

$$[p] = -M^2\left[\frac{1}{\rho}\right], \tag{22.11}$$

again using $[M] = 0$.

22.3   Treating the Jump Conditions

Since the normal velocity is discontinuous across the interface, caution is needed in applying numerical discretizations near the interface. For example, when discretizing the unreacted fluid velocity near the interface, one should avoid using values of the reacted fluid velocity. Following the ghost fluid methodology, a band of ghost cells on the reacted side of the interface is populated with unreacted ghost velocities that can be used in the discretization of the unreacted fluid velocity. This is done using equation (22.9) to obtain

$$u_u^G = u_r - M \left(\frac{1}{\rho_r} - \frac{1}{\rho_u} \right) n_1, \tag{22.12}$$

$$v_u^G = v_r - M \left(\frac{1}{\rho_r} - \frac{1}{\rho_u} \right) n_2, \tag{22.13}$$

$$w_u^G = w_r - M \left(\frac{1}{\rho_r} - \frac{1}{\rho_u} \right) n_3, \tag{22.14}$$

where $\vec{N} = (n_1, n_2, n_3)$ is the local unit normal. Similarly, reacted ghost velocities are defined on a band of ghost cells on the unreacted side of the interface and used in the discretization of the reacted fluid velocity.

When solving equation (18.23),

$$\nabla \cdot \left(\frac{\nabla p^\star}{\rho} \right) = \nabla \cdot \vec{V}^\star, \tag{22.15}$$

for the scaled pressure $p^\star$, the jump in pressure given by equation 22.11 as

$$[p^\star] = -\Delta t M^2 \left(\frac{1}{\rho_r} - \frac{1}{\rho_u} \right) \tag{22.16}$$

is accounted for using the techniques developed in Liu et al. [106] and Kang et al. [91].

Figure 22.1 shows the time evolution of two initially circular flame fronts as they grow to merge together. Figure 22.2 shows a snapshot of the velocity field, illustrating its discontinuous nature across the interface. Figure 22.3 shows the time evolution of two initially spherical flame fronts as they grow to merge together.

Recently, Nguyen et al. [118] extended this approach to model fire for computer graphics. In Figure 22.4, the $\phi = 0$ isocontour is used to render a typical blue flame core. S is smaller for the larger blue core on the right. Figure 22.5 illustrates the effect of increased expansion as the density jump is increased from left to right. The yellow flame color is calculated using a blackbody radiation model based on the temperature profile of the hot gas emitted at the flame front. Figures 22.6 shows a ball catching on fire, and Figure 22.7 shows a campfire.

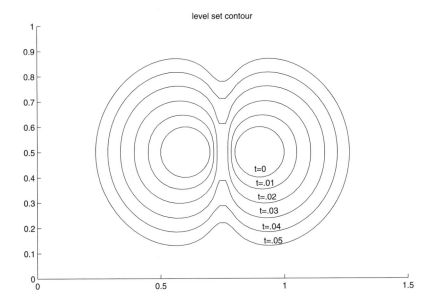

Figure 22.1. Time evolution of two initially circular flame fronts as they grow to merge together.

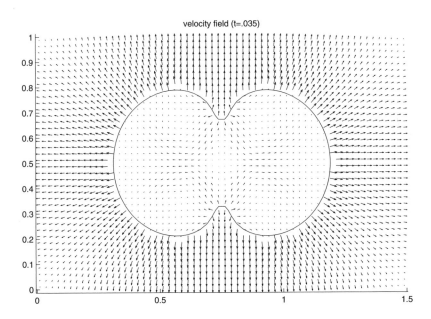

Figure 22.2. Discontinuous velocity field depicted shortly after the two flame fronts merge.

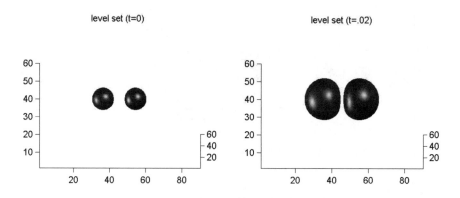

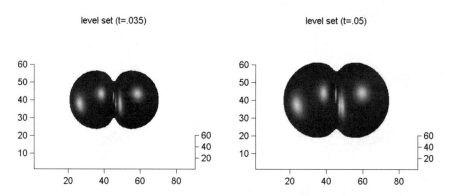

Figure 22.3. Time evolution of two initially spherical flame fronts as they grow to merge together.

Figure 22.4. Typical blue cores rendered using the zero isocontour of the level set function. (See also color figure, Plate 21.)

Figure 22.5. The density ratio of the unburnt to burnt gas is increased from left to right, illustrating the effect of increased expansion. (See also color figure, Plate 22.)

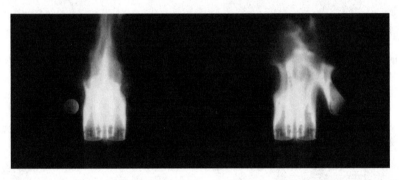

Figure 22.6. A flammable ball catches on fire as it passes through a flame. (See also color figure, Plate 23.)

Figure 22.7. Campfire with realistic lighting of the surrounding rocks. (See also color figure, Plate 24.)

23
Heat Flow

23.1 Heat Equation

Starting from conservation of mass, momentum, and energy one can derive

$$\rho e_t + \rho \vec{V} \cdot \nabla e + p \nabla \cdot \vec{V} = \nabla \cdot (k \nabla T), \qquad (23.1)$$

where k is the thermal conductivity and T is the temperature. Assuming that e depends on at most temperature, and that the specific heat at constant volume, c_v is constant leads to $e = e_o + c_v(T - T_o)$, where e_o is the internal energy per unit mass at some reference temperature T_o (see, for example, Atkins [10]). This and the incompressibility assumption $\nabla \cdot \vec{V} = 0$ simplify equation (23.1) to

$$\rho c_v T_t + \rho c_v \vec{V} \cdot \nabla T = \nabla \cdot (k \nabla T), \qquad (23.2)$$

which can be further simplified to the standard heat equation

$$\rho c_v T_t = \nabla \cdot (k \nabla T) \qquad (23.3)$$

by ignoring the effects of convection, i.e., setting $\vec{V} = 0$.

Applying explicit Euler time discretization to equation (23.3) results in

$$\frac{T^{n+1} - T^n}{\Delta t} = \frac{1}{\rho c_v} \nabla \cdot (k \nabla T^n), \qquad (23.4)$$

where either Dirichlet or Neumann boundary conditions can be applied on the boundaries of the computational domain. Assuming that ρ and c_v are constants allows us to rewrite this equation as

$$\frac{T^{n+1} - T^n}{\Delta t} = \nabla \cdot \left(\hat{k} \nabla T^n \right) \tag{23.5}$$

with $\hat{k} = k/(\rho c_v)$. Standard central differencing can be used for the spatial derivatives and a time step restriction of

$$\Delta t \hat{k} \left(\frac{2}{(\Delta x)^2} + \frac{2}{(\Delta y)^2} + \frac{2}{(\Delta z)^2} \right) \leq 1 \tag{23.6}$$

is needed for stability.

Implicit Euler time discretization

$$\frac{T^{n+1} - T^n}{\Delta t} = \nabla \cdot \left(\hat{k} \nabla T^{n+1} \right) \tag{23.7}$$

avoids this time step stability restriction. This equation can be rewritten as

$$T^{n+1} - \Delta t \nabla \cdot \left(\hat{k} \nabla T^{n+1} \right) = T^n \tag{23.8}$$

where the $\nabla \cdot \left(\hat{k} \nabla T^{n+1} \right)$ term is discretized using central differencing. For each unknown T_i^{n+1}, equation (23.8) is used to fill in one row of a matrix, creating a linear system of equations. Since the resulting matrix is symmetric, a number of fast linear solvers can be used (e.g., a PCG method with an incomplete Choleski preconditioner; see Golub and Van Loan [75]). Equation (23.7) is first-order accurate in time and second-order accurate in space, and Δt needs to be chosen proportional to Δx^2, in order to obtain an overall asymptotic accuracy of $O(\Delta x^2)$. However, the stability of the implicit Euler method allows one to chose Δt proportional to Δx saving dramatically on CPU time. The Crank-Nicolson scheme

$$\frac{T^{n+1} - T^n}{\Delta t} = \frac{1}{2} \nabla \cdot \left(\hat{k} \nabla T^{n+1} \right) + \frac{1}{2} \nabla \cdot \left(\hat{k} \nabla T^n \right) \tag{23.9}$$

can be used to achieve second-order accuracy in both space and time with Δt proportional to Δx. For the Crank-Nicolson scheme,

$$T^{n+1} - \frac{\Delta t}{2} \nabla \cdot \left(\hat{k} \nabla T^{n+1} \right) = T^n + \frac{\Delta t}{2} \nabla \cdot \left(\hat{k} \nabla T^n \right) \tag{23.10}$$

is used to create a symmetric linear system of equations for the unknowns T_i^{n+1}. Again, all spatial derivatives are computed using standard central differencing.

23.2   Irregular Domains

Instead of a uniform Cartesian domain, suppose we wish to solve equation (23.3) on an irregularly shaped domain, for example in the interior of the two-dimensional outline depicted in Figure (23.1). If one takes the rather simple approach of embedding this complicated domain in a uniform Cartesian grid, then a level set function can be used to define the boundary of the irregular region. The heat equation 23.3 can then be solved with, for example, Dirichlet $T = g(\vec{x}, t)$ boundary conditions applied to the boundary where $\phi = 0$. More complicated boundary conditions can be used as well.

The spatial derivatives are computed with the aid of the given values of $T = g(\vec{x}, t)$ on the interface. When using explicit Euler time discretization, the time-step restriction needed for stability becomes

$$\triangle t \hat{k} \left(\frac{2}{(\theta_1 \triangle x)^2} + \frac{2}{(\theta_2 \triangle y)^2} + \frac{2}{(\theta_3 \triangle z)^2} \right) \leq 1, \qquad (23.11)$$

where θ_1, θ_2, and θ_3 are the cell fractions in each spatial dimension for cells cut by the interface with $0 < \theta_i \leq 1$. Since the θ_i's can be arbitrarily small, leading to arbitrarily small time steps, implicit methods need to be used, e.g., backward Euler or Crank-Nicolson. Then a linear system of equations can be solved for the unknowns T_i^{n+1}. Since the coefficient matrix depends on the details of the spatial discretization, a robust method for treating the cut cells is crucial. Below, we outline how to do this for the simpler variable-coefficient Poisson equation, which has spatial derivatives identical to those of the heat equation.

23.3 Poisson Equation

Consider the variable-coefficient Poisson equation

$$\nabla \cdot (\beta(\vec{x}) \nabla u(\vec{x})) = f(\vec{x}), \qquad (23.12)$$

where $\beta(\vec{x})$ is positive and bounded below by some $\epsilon > 0$. As above, consider an irregularly shaped domain (as in Figure 23.1) defined by a level set function on a Cartesian grid with Dirichlet $u = g(\vec{x}, t)$ boundary conditions on the $\phi = 0$ isocontour.

For simplicity consider the one-dimensional case $(\beta u_x)_x = f$. Since β and ϕ are known only at the grid nodes, their values between grid nodes are defined by the linear average of the nodal values, e.g., $\beta_{i+\frac{1}{2}} = (\beta_i + \beta_{i+1})/2$. In the absence of cut cells, the standard discretization

$$\frac{\beta_{i+\frac{1}{2}} \left(\frac{u_{i+1} - u_i}{\triangle x} \right) - \beta_{i-\frac{1}{2}} \left(\frac{u_i - u_{i-1}}{\triangle x} \right)}{\triangle x} = f_i \qquad (23.13)$$

can be used to solve this problem. For each unknown u_i, equation (23.13) is used to fill in one row of a matrix, creating a linear system of equations. The resulting matrix is symmetric and can be solved with a number of fast linear solvers.

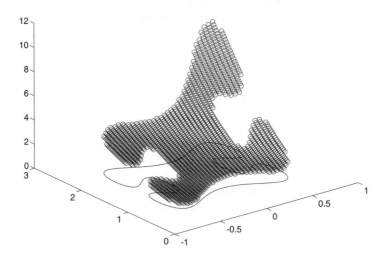

Figure 23.1. Solution of the two-dimensional Poisson equation $\nabla \cdot (\beta \nabla u) = f$ with Dirichlet boundary conditions. The circles are the computed solution, and the solid line contour outlines the irregularly shaped computational domain.

Suppose that an interface point x_I is located between two grid points x_i and x_{i+1} with a Dirichlet $u = u_I$ boundary condition applied at x_I. Consider computing the numerical solution in the domain to the left of x_I. Equation (23.13) is valid for all the unknowns to the left and including u_{i-1}, but can no longer be applied at x_i to solve for u_i, since the subdomain to the left of x_I does not contain a valid value of u_{i+1}. This can be remedied by defining a ghost value of u_{i+1}^G at x_{i+1} and rewriting equation (23.13) as

$$\frac{\beta_{i+\frac{1}{2}} \left(\frac{u_{i+1}^G - u_i}{\triangle x} \right) - \beta_{i-\frac{1}{2}} \left(\frac{u_i - u_{i-1}}{\triangle x} \right)}{\triangle x} = f_i \qquad (23.14)$$

in order to solve for u_i. Possible candidates for u_{i+1}^G include

$$u_{i+1}^G = u_I \qquad (23.15)$$

$$u_{i+1}^G = \frac{u_I + (\theta - 1) \, u_i}{\theta} \qquad (23.16)$$

and

$$u_{i+1}^G = \frac{2u_I + \left(2\theta^2 - 2\right) u_i + \left(-\theta^2 + 1\right) u_{i-1}}{\theta^2 + \theta} \qquad (23.17)$$

with constant, linear, and quadratic extrapolation respectively. Here $\theta \in [0, 1]$ is defined by $\theta = (x_I - x_i)/\triangle x$, and it can be calculated as $\theta = |\phi|/\triangle x$, since ϕ is a signed distance function vanishing at x_I. Since equations (23.16) and (23.17) are poorly behaved for small θ, they are not used when $\theta \leq \triangle x$. Instead, u_i is set equal to u_I, which effectively moves the interface location from x_I to x_i. This second-order accurate perturbation of the interface location does not degrade the overall second-order accuracy of the solution obtained using equation (23.13) to solve for the remaining unknowns. Furthermore, $u_i = u_I$ is second-order accurate as long as the solution has bounded first derivatives.

Plugging equation (23.17) into equation (23.13) gives an asymmetric discretization of

$$\frac{\left(\frac{u_I - u_i}{\theta \triangle x}\right) - \left(\frac{u_i - u_{i-1}}{\triangle x}\right)}{.5\left(\theta \triangle x + \triangle x\right)} = f_i \qquad (23.18)$$

(when $\beta = 1$). Equation (23.18) is the asymmetric discretization used by Chen et al. [43] to obtain second-order accurate numerical methods in the context of solving Stefan problems. Alternatively, Gibou et al. [44] pointed out that plugging equation (23.16) into equation (23.13) gives a symmetric discretization of

$$\frac{\beta_{i+\frac{1}{2}}\left(\frac{u_I - u_i}{\theta \triangle x}\right) - \beta_{i-\frac{1}{2}}\left(\frac{u_i - u_{i-1}}{\triangle x}\right)}{\triangle x} = f_i \qquad (23.19)$$

based on linear extrapolation in the cut cell. It turns out that this symmetric discretization is second-order accurate as well. Moreover, since the discretization is symmetric, the linear system of equations can be solved with a number of fast methods such as PCG with an incomplete Choleski preconditioner.

To see this, assume that the standard second-order accurate discretization in equation (23.13) is used to obtain the standard linear system of equations for u at every grid node except x_i, and equation (23.14) is used to write a linear equation for u_i, introducing a new unknown u_{i+1}^G. The system is closed with equation (23.16) for u_{i+1}^G. In practice, equations (23.16) and (23.14) are combined to obtain equation (23.19) and a symmetric linear system. Solving this linear system of equations leads to well-determined values of u at each grid node in the subdomain as well as a well-determined value of u_{i+1}^G (from equation (23.16)). Designate $\vec{u}$ as the solution vector containing all these values of u.

Next, consider a modified problem where a Dirichlet boundary condition of $u_{i+1} = u_{i+1}^G$ is specified at x_{i+1} with u_{i+1}^G chosen to be the value of u_{i+1}^G from $\vec{u}$ (defined above). This modified problem can be discretized to second-order accuracy everywhere using the standard discretization in equation (23.13) at every node except at x_i, where equation (23.14) is used. Note that equation (23.14) *is* the standard second-order accurate discretization when a Dirichlet boundary condition of $u_{i+1} = u_{i+1}^G$ is applied at x_{i+1}. Thus, this new linear system of equations can be solved in standard fashion to obtain a second-order accurate solution at every grid node. The realization that $\vec{u}$ (defined above) is an exact solution to *this* new linear system implies that $\vec{u}$ is a valid second-order accurate solution to this modified problem.

Since $\vec{u}$ is a second-order accurate solution to the modified problem, $\vec{u}$ can be used to obtain the interface location for the modified problem to second-order accuracy. The linear interpolant that uses u_i at x_i and u_{i+1}^G at x_{i+1} predicts an interface location of *exactly* x_I. Since higher-order accurate interpolants (higher than linear) can contribute at most an $O(\Delta x^2)$ perturbation of the predicted interface location the interface location dictated by the modified problem is at most an $O(\Delta x^2)$ perturbation of the true interface location, x_I. Thus, $\vec{u}$ is a second-order accurate solution to a modified problem where the interface location has been perturbed by $O(\Delta x^2)$. This makes $\vec{u}$ a second-order accurate solution to the original problem as well.

Note that plugging equation (23.15) into equation (23.13) effectively perturbs the interface location by an $O(\Delta x)$ amount, resulting in a first-order accurate algorithm.

When β is spatially varying, $\beta_{i+1/2}$ in equation (23.19) can be determined from a ghost value β_{i+1}^G and the usual averaging $\beta_{i+1/2} = \left(\beta_i + \beta_{i+1}^G\right)/2$, where the ghost value is defined using linear extrapolation

$$\beta_{i+1}^G = \frac{\beta_I + (\theta - 1)\beta_i}{\theta} \tag{23.20}$$

according to equation (23.16).

In multiple spatial dimensions, the equations are discretized in a dimension-by-dimension manner using the one-dimensional discretization outlined above independently on $(\beta u_x)_x$, $(\beta u_y)_y$, and $(\beta u_z)_z$. Figure 23.1 shows a typical solution obtained in two spatial dimensions with a spatially varying β.

The same techniques can be used to discretize the spatial terms in equation (23.8) or (23.10) to obtain symmetric linear systems of equations for the unknown temperatures T_i^{n+1}. Again, the symmetry allows us to exploit a number of fast solvers such as PCG.

23.4   Stefan Problems

Stefan problems model interfaces across which an unreacted incompressible material is converted into a reacted incompressible material. The interface velocity is $W = D\vec{N}$, where $D = (V_N)_u + S$ for some reaction speed S. Here the "u" subscript denotes an unreacted material quantity. Including the effects of thermal conductivity, the Rankine-Hugoniot jump condition for conservation of energy is

$$\left[\left(\rho e + \frac{\rho(V_N - D)^2}{2} + p \right) (V_N - D) \right] = \left[k\nabla T \cdot \vec{N} \right], \qquad (23.21)$$

where we have assumed that $D \neq V_N$ (i.e., $S \neq 0$), so that the tangential velocities are continuous across the interface. This equation can be rewritten as

$$-\rho_u S \left(\left[e + \frac{p}{\rho} \right] + \frac{\rho_u^2 S^2}{2} \left[\frac{1}{\rho^2} \right] \right) = \left[k\nabla T \cdot \vec{N} \right] \qquad (23.22)$$

using the Rankine-Hugoniot jump condition for conservation of mass, $[\rho(V_N - D)] = 0$. Assuming that the enthalpy per unit mass $h = e + (p/\rho)$ depends on at most temperature, and that the specific heat at constant pressure c_p is constant leads to $h = h_o + c_p (T - T_o)$, where h_o is the enthalpy per unit mass at some reference temperature T_o; see [10]. This allows us to rewrite equation (23.22) as

$$-\rho_u S \left([h_o] + [c_p] (T_I - T_o) + \frac{\rho_u^2 S^2}{2} \left[\frac{1}{\rho^2} \right] \right) = \left[k\nabla T \cdot \vec{N} \right], \qquad (23.23)$$

where we have used the fact that the temperature is continuous across the interface, $[T] = 0$, and labeled the interface temperature T_I. It is convenient to choose the reference temperature T_o equal to the standard temperature at which the reaction takes place; e.g., in the case of freezing water $T_o = 273\ K$.

For the Stefan problem we assume that there is no expansion across the front (i.e., $[\rho] = 0$), reducing the Rankine-Hugoniot jump conditions for mass and momentum to $[V_N] = 0$ and $[p] = 0$, respectively. Then equation (23.23) reduces to

$$-\rho S \left([h_o] + [c_p] (T_I - T_o) \right) = \left[k\nabla T \cdot \vec{N} \right], \qquad (23.24)$$

where $\rho = \rho_u = \rho_r$. Finally, the standard interface boundary condition of $T_I = T_o$ reduces this last equation to

$$-\rho S [h_o] = \left[k\nabla T \cdot \vec{N} \right], \qquad (23.25)$$

where $[h_o]$ is calculated at the reaction temperature of $T_I = T_o$.

The Stefan problem is generally solved in three steps. First, the interface velocity is determined using equation (23.25). This is done by first computing $T_N = \nabla T \cdot \vec{N}$ in a band about the interface, and then extrapolating these values across the interface (see equation (8.1)) so that both $(T_N)_u$ and $(T_N)_r$ are defined at every grid point in a band about the interface, allowing the reaction speed S to be computed in a node-by-node fashion.

Next, the level set method is used to evolve the interface to its new location. Finally, the temperature is calculated in each subdomain using a Dirichlet boundary condition on the temperature at the interface. This Dirichlet boundary condition decouples the problem into two disjoint subproblems that can each be solved separately using the techniques described earlier in this chapter for the heat equation. For more details, see [44].

Figure 23.2 shows a sample calculation of an outwardly growing interface in three spatial dimensions. Figure 23.3 shows two-dimensional results obtained using anisotropic surface tension. The interface condition is the fourfold anisotropy boundary condition

$$T = -0.001 \left(\frac{8}{3} \right) \sin^4 \left(2(\theta - \theta_0) \right) \kappa$$

with (left) $\theta_0 = 0$ and (right) $\theta_0 = \pi/4$. The shape of the crystal in the right figure is that of the crystal in the left figure rotated by $\pi/4$, demonstrating that the artificial grid anisotropy is negligible.

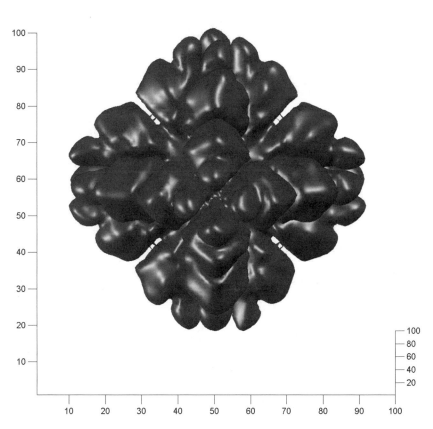

Figure 23.2. Stefan problem in three spatial dimensions. A supercooled material in the exterior region promotes unstable growth.

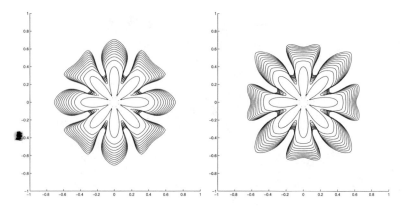

Figure 23.3. Grid orientation effects with anisotropic surface tension. The interface condition is the fourfold anisotropy boundary condition $T = -0.001 \left(\frac{8}{3}\right) \sin^4\left(2(\theta - \theta_0)\right) \kappa$ with (left) $\theta_0 = 0$ and (right) $\theta_0 = \pi/4$. The shape of the crystal in the right figure is that of the crystal in the left figure rotated by $\pi/4$, demonstrating that the artificial grid anisotropy is negligible.

References

[1] Adalsteinsson, D. and Sethian, J., *The Fast Construction of Extension Velocities in Level Set Methods*, J. Comput. Phys. 148, 2–22 (1999).

[2] Adalsteinsson, D. and Sethian, J., *A Fast Level Set Method for Propagating Interfaces*, J. Comput. Phys. 118, 269-277 (1995).

[3] Aivazis, M., Goddard, W., Meiron, D., Ortiz, M., Pool, J., and Shepherd, J., *A Virtual Test Facility for Simulating the Dynamic Response of Materials*, Comput. in Sci. and Eng. 2, 42–53 (2000).

[4] Alouges, F., *A New Algorithm for Computing Liquid Crystal Stable Configurations: The Harmonic Map Case*, SIAM J. Num. Anal. 34, 1708–1726 (1997).

[5] Alvarez, L., Guichard, F., Lions, P.-L., and Morel, J.-M., *Axioms and Fundamental Equations of Image Processing*, Arch. Rat. Mech. and Analys. 16, 200–257 (1993).

[6] Amenta, N. and Bern, M., *Surface Reconstruction by Voronoi Filtering*, Discrete and Comput. Geometry 22, 481–504 (1999).

[7] Anderson, J., *Computational Fluid Dynamics*, McGraw Hill Inc. 1995.

[8] Anderson, D., Tannehill, J., and Pletcher, R., *Computational Fluid Mechanics and Heat Transfer*, Hemisphere Publishing Corporation (1984).

[9] Aslam, T., *A Level-Set Algorithm for Tracking Discontinuities in Hyperbolic Conservation Laws, I. Scalar Equations*, J. Comput. Phys. 167, 413–438 (2001).

[10] Atkins, P., *Physical Chemistry*, 5th edition, Freeman, (1994).

[11] Batchelor, G., *An Introduction to Fluid Dynamics*, Cambridge University Press (1967).

[12] Bardi, M. and Osher, S., *The Nonconvex Multidimensional Riemann Problem for Hamilton–Jacobi Equations*, SIAM J. Math. Anal. 22, 344–351 (1991).

[13] Bell, J., Colella, P., and Glaz, H., *A Second Order Projection Method for the Incompressible Navier–Stokes Equations*, J. Comput. Phys. 85, 257–283 (1989).

[14] Benson, D., *Computational Methods in Lagrangian and Eulerian Hydrocodes*, Computer Methods in App. Mech. and Eng. 99, 235–394 (1992).

[15] Benson, D., *A New Two-Dimensional Flux-Limited Shock Viscosity for Impact Calculations*, Computer Methods in App. Mech. and Eng. 93, 39–95 (1991).

[16] Bloomenthal, J., Bajaj, C., Blinn, J., Cani-Gascuel, M.-P., Rockwood, A., Wyvill, B., and Wyvill, G., *Introduction to Implicit Surfaces*, Morgan Kaufmann Publishers Inc., San Francisco (1997).

[17] Boissonat, J. and Cazals, F., *Smooth Shape Reconstruction via Natural Neighbor Interpolation of Distance Functions*, ACM Symposium on Comput. Geometry (2000).

[18] Brackbill, J., Kothe, D., and Zemach, C., *A Continuum Method for Modeling Surface Tension*, J. Comput. Phys. 100, 335–354 (1992).

[19] Brown, D., Cortez, R., and Minion, M., *Accurate Projection Methods for the Incompressible Navier–Stokes Equations*, J. Comput. Phys. 168, 464–499 (2001).

[20] Bruckstein, A., *On Shape from Shading*, Comput. Vision Graphics Image Process. 44, 139–154 (1988).

[21] Burchard, P., Chen, S., Osher, S., and Zhao, H.-K., Level Set Systems Report, 11/8/2000.

[22] Burchard, P., Cheng, L.-T., Merriman, B., and Osher, S., *Motion of Curves in Three Spatial Dimensions Using a Level Set Approach*, J. Comput. Phys. 170, 720–741 (2001).

[23] Caiden, R., Fedkiw, R., and Anderson, C., *A Numerical Method for Two-Phase Flow Consisting of Separate Compressible and Incompressible Regions*, J. Comput. Phys. 166, 1–27 (2001).

[24] Caramana, E., Burton, D., Shashkov, M., and Whalen, P., *The Construction of Compatible Hydrodynamics Algorithms Utilizing Conservation of Total Energy*, J. Comput. Phys. 146, 227–262 (1998).

[25] Caramana, E., Rousculp, C., and Burton, D., *A Compatible, Energy and Symmetry Preserving Lagrangian Hydrodynamics Algorithm in Three-Dimensional Cartesian Geometry*, J. Comput. Phys. 157, 89–119 (2000).

[26] Carr, J., Beatson, R., Cherrie, J., Mitchell, T., Fright, W., McCallum, B., and Evans, T., *Reconstruction and Representation of 3D Objects with Radial Basis Functions*, SIGGRAPH '01, 67–76 (2001).

[27] Caselles, V., Catté, F., Coll, T., and Dibos, F., *A Geometric Model for Active Contours in Image Processing*, Numerische Mathematik 66, 1–31 (1993).

[28] Caselles, V., Kimmel, R., and Sapiro, G., *Geodesic Active Contours*, Int. J. Comput. Vision 22, 61–79 (1997).

[29] Caselles, V., Morel, J.-M., Sapiro, G., and Tannenbaum, A., eds., *Special Issue on Partial Differential Equations and Geometry-Driven Diffusion in Image Processing and Analysis*, IEEE Transactions on Image Processing 7, 269–473 (1998).

[30] Chan, T., Golub, G., and Mulet, P., *A Nonlinear Primal–Dual Method for Total Variation Based Image Restoration*, SIAM J. Sci. Comput. 15, 892–915 (1994).

[31] Chan, T., Osher, S., and Shen, J., *The Digital TV Filter and Nonlinear Denoising*, IEEE Trans. on Image Processing 10, 231–241 (2001).

[32] Chan, T., Sandberg, B., and Vese, L., *Active Contours without Edges for Vector-Valued Images*, J. Visual Commun. and Image Rep. 11, 130–141 (2000).

[33] Chan, T. and Vese, L., *Active Contour and Segmentation Models using Geometric PDE's for Medical Imaging*, in "Geometric Methods in Bio-Medical", R. Malladi (editor), Mathematics and Visualization, Springer (March 2002).

[34] Chan, T. and Vese, L., *Active Contours without Edges*, IEEE Trans. on Image Processing 10, 266–277 (2001).

[35] Chan, T. and Vese, L., *An Active Contour Model without Edges*, "Scale-Space Theories in Computer Vision," Lect. Notes in Comput. Sci. 1682, 141–151 (1999).

[36] Chan, T. and Vese, L., *Image Segmentation using Level Sets and the Piecewise-Constant Mumford–Shah Model*, UCLA CAM Report 00–14 (2000).

[37] Chan, T. and Vese, L., *A Level Set Algorithm for Minimizing the Mumford–Shah Functional in Image Processing*, in IEEE/Comput. Soc. Proc. of the 1st IEEE Workshop on "Variational and Level Set Methods in Computer Vision", 161–168 (2001).

[38] Chang, Y., Hou, T., Merriman, B., and Osher, S., *A Level Set Formulation of Eulerian Interface Capturing Methods for Incompressible Fluid Flows*, J. Comput. Phys. 124, 449–464 (1996).

[39] Chen, S., Johnson, D., and Raad, P., *Velocity Boundary Conditions for the Simulation of Free Surface Fluid Flow*, J. Comput. Phys. 116, 262–276 (1995).

[40] Chen, S., Johnson, D., Raad, P., and Fadda, D., *The Surface Marker and Micro Cell Method*, Int. J. for Num. Meth. in Fluids 25, 749–778 (1997).

[41] Catte, F., Lions, P.-L., Morel, J.-M., and Coll, T., *Image Selective Smoothing and Edge Detection by Nonlinear Diffusion*, SIAM J. Num. Anal. 29, 182–193 (1992).

[42] Chambolle, A. and Lions, P.-L., *Image Recovery via Total Variation Minimization and Related Problems*, Numer. Math. 76, 167–188 (1997).

[43] Chen, S., Merriman, B., Osher, S., and Smereka, P., *A simple level set method for solving Stefan problems*, J. Comput. Phys. 135, 8–29 (1997).

[44] Gibou, F., Fedkiw, R., Cheng, L.-T., and Kang, M., *A Second Order Accurate Symmetric Discretization of the Poisson Equation on Irregular Domains*, J. Comput. Phys. V. 176, pp 1–23 (2002).

[45] Chern, I.-L. and Colella, P. *A Conservative Front Tracking Method for Hyperbolic Conservation Laws*, UCRL JC-97200 LLNL (1987).

[46] Chorin, A., *A Numerical Method for Solving Incompressible Viscous Flow Problems*, J. Comput. Phys. 2, 12–26 (1967).

[47] Chorin, A., *Numerical Solution of the Navier–Stokes Equations*, Math. Comp. 22, 745–762 (1968).

[48] Chopp, D., *Computing Minimal Surfaces via Level Set Curvature Flow*, J. Comput. Phys. 106, 77–91 (1993).

[49] Chopp, D., *Some Improvements of the Fast Marching Method*, SIAM J. Sci. Comput. 223, pp. 230–244 (2001).

[50] Colella, P., Majda, A., and Roytburd, V., *Theoretical and Numerical Structure for Reacting Shock Waves*, SIAM J. Sci. Stat. Comput. 7, 1059–1080 (1986).

[51] Courant, R., Issacson, E., and Rees, M., *On the Solution of Nonlinear Hyperbolic Differential Equations by Finite Differences*, Comm. Pure and Applied Math 5, 243–255 (1952).

[52] Crandall, M. and Lions, P.-L., *Viscosity Solutions of Hamilton–Jacobi Equations*, Trans. Amer. Math. Soc. 277, 1–42 (1983).

[53] Crandall, M. and Lions, P.-L., *Two Approximations of Solutions of Hamilton–Jacobi Equations*, Math. Comput. 43, 1–19 (1984).

[54] Davis, W., *Equation of state for detonation products*, Proceedings Eighth Symposium on Detonation, 785–795 (1985).

[55] Donat, R., and Marquina, A., *Capturing Shock Reflections: An Improved Flux Formula*, J. Comput. Phys. 25, 42–58 (1996).

[56] E, W. and Liu, J.-G., *Finite Difference Schemes for Incompressible Flows in the Velocity-Impulse Density Formulation*, J. Comput. Phys. 130, 67–76 (1997).

[57] E, W. and Wang, X.-P., *Numerical Methods for the Landau–Lifshitz Equation*, SIAM J. Num. Anal. 381, 1647–1665 (2000).

[58] Edelsbrunner, H., *Shape Reconstruction with Delaunay Complex*, Proc. of Latin '98, Theoretical Informatics 1380, Lect. Notes in CS, 119–132, Springer-Verlag (1998).

[59] Engquist, B., Fatemi, E., and Osher, S., *Numerical Solution of the High Frequency Asymptotic Expansion of the Scalar Wave Equation*, J. Comput. Phys. 120, 145–155 (1995).

[60] Engquist, B., Runborg, O., and Tornberg, A.-K., *High Frequency Wave Propagation by the Segment Projection Method*, J. Comput. Phys (in press).

[61] Enright, D., Fedkiw, R., Ferziger, J., and Mitchell, I., *A Hybrid Particle Level Set Method for Improved Interface Capturing*, J. Comput. Phys. (in press).

[62] Fedkiw, R., *Coupling an Eulerian Fluid Calculation to a Lagrangian Solid Calculation with the Ghost Fluid Method*, J. Comput. Phys. 175, 200–224 (2002).

[63] Fedkiw, R., Aslam, T., Merriman, B., and Osher, S., *A Non-Oscillatory Eulerian Approach to Interfaces in Multimaterial Flows (The Ghost Fluid Method)*, J. Comput. Phys. 152, 457–492 (1999).

[64] Fedkiw, R., Aslam, T., and Xu, S., *The Ghost Fluid Method for Deflagration and Detonation Discontinuities*, J. Comput. Phys. 154, 393–427 (1999).

[65] Fedkiw, R., Merriman, B., Donat, R. and Osher, S., *The Penultimate Scheme for Systems of Conservation Laws: Finite Difference ENO with Marquina's Flux Splitting*, Progress in Numerical Solutions of Partial Differential Equations, Arcachon, France, edited by M. Hafez (July 1998).

[66] Fedkiw, R., Marquina, A., and Merriman, B., *An Isobaric Fix for the Overheating Problem in Multimaterial Compressible Flows*, J. Comput. Phys. 148, 545–578 (1999).

[67] Fedkiw, R., Merriman, B., and Osher, S., *Efficient Characteristic Projection in Upwind Difference Schemes for Hyperbolic Systems (The Complementary Projection Method)*, J. Comput. Phys. 141, 22–36 (1998).

[68] Fedkiw, R., Merriman, B., and Osher, S., *Numerical Methods for a Mixture of Thermally Perfect and/or Calorically Perfect Gaseous Species with Chemical Reactions*, J. Comput. Phys. 132, 175–190 (1997).

[69] Fedkiw, R., Merriman, B., and Osher, S., *Simplified Upwind Discretization of Systems of Hyperbolic Conservation Laws Containing Advection Equations*, J. Comput. Phys. 157, 302–326 (2000).

[70] Fedkiw, R., Stam, J., and Jensen, H., *Visual Simulation of Smoke*, Siggraph 2001 Annual Conference, 23–30 (2001).

[71] Foster, N. and Fedkiw, R., *Practical Animation of Liquids*, Siggraph 2001 Annual Conference, 15–22 (2001).

[72] Glimm, J., Grove, J., Li, X., and Zhao, N., *Simple front tracking*, Contemporary Math. 238, 133–149 (1999).

[73] Glimm, J., Marchesin, D., and McBryan, O., *Subgrid resolution of fluid discontinuities, II.*, J. Comput. Phys. 37, 336–354 (1980).

[74] Godunov, S.K., *A Finite Difference Method for the Computation of Discontinuous Solutions of the Equations of Fluid Dynamics*, Mat. Sb. 47, 357–393 (1959).

[75] Golub, G. and Van Loan, C., *Matrix Computations*, The Johns Hopkins University Press, Baltimore (1989).

[76] Greengard, L. and Rokhlin, V. *A Fast Algorithm for Particle Simulations*, J. Comput. Phys. 73, 325–348 (1987).

[77] Guichard, F. and Morel, J.-M., *Image Iterative Smoothing and P.D.E's*, Notes de Cours du Centre Emile Borel, Institut Henri Poincaré (1998).

[78] Hancock, S., *PISCES 2DELK Theoretical Manual*, Physics International (1985).

[79] Harlow, F. and Welch, J., *Numerical Calculation of Time-Dependent Viscous Incompressible Flow of Fluid with a Free Surface*, The Physics of Fluids 8, 2182–2189 (1965).

[80] Harten, A., *High Resolution Schemes for Hypersonic Conservation Laws*, J. Comput. Phys. 49, 357–393 (1983).

[81] Harten, A., Engquist, B., Osher, S., and Chakravarthy, S., *Uniformly High-Order Accurate Essentially Non-Oscillatory Schemes III*, J. Comput. Phys. 71, 231–303 (1987).

[82] Heath, M., *Scientific Computing*, The McGraw-Hill Companies Inc. (1997).

[83] Helenbrook, B. and Law, C., *The Role of Landau–Darrieus Instability in Large Scale Flows*, Combustion and Flame 117, 155–169 (1999).

[84] Helenbrook, B., Martinelli, L., and Law, C., *A Numerical Method for Solving Incompressible Flow Problems with a Surface of Discontinuity*, J. Comput. Phys. 148, 366–396 (1999).

[85] Helmsen, J., Puckett, E., Colella, P., and Dorr, M., *Two New Methods for Simulating Photolithography Development in 3D*, Proc. SPIE 2726, 253–261 (1996).

[86] Hirsch, C., *Numerical Computation of Internal and External Flows, Volume 1: Fundamentals of Numerical Discretization*, John Wiley and Sons Ltd. (1988).

[87] Hirsch, C., *Numerical Computation of Internal and External Flows, Volume 2: Computational Methods for Inviscid and Viscous Flows*, John Wiley and Sons Ltd. (1990).

[88] Jiang, G.-S. and Peng, D., *Weighted ENO Schemes for Hamilton Jacobi Equations*, SIAM J. Sci. Comput. 21, 2126–2143 (2000).

[89] Jiang, G.-S. and Shu, C.-W., *Efficient Implementation of Weighted ENO Schemes*, J. Comput. Phys. 126, 202–228 (1996).

[90] Juric, D. and Tryggvason, G., *Computations of Boiling Flows*, Int. J. Multiphase Flow 24, 387–410 (1998).

[91] Kang, M., Fedkiw, R., and Liu, X.-D., *A Boundary Condition Capturing Method for Multiphase Incompressible Flow*, J. Sci. Comput. 15, 323–360 (2000).

[92] Karni, S., *Hybrid multifluid algorithms*, SIAM J. Sci. Comput. 17, 1019–1039 (1996).

[93] Karni, S., *Multicomponent Flow Calculations by a Consistent Primitive Algorithm*, J. Comput. Phys. 112, 31–43 (1994).

[94] Kass, M., Witkin, A., and Terzopoulos, D., *Snakes: Active Contour Models*, Int. J. of Comp. Vision 1, 321–331 (1988).

[95] Kichenassamy, S., Kumar, A., Olver, P., Tannenbaum, A., Yezzi, A., *Conformal Curvature Flows: From Phase Transitions to Active Vision*, Archive for Rational Mech. and Anal. 134, 275–301 (1996).

[96] Kim, J. and Moin, P., *Application of a Fractional-Step Method to Incompressible Navier–Stokes Equations*, J. Comput. Phys. 59, 308–323 (1985).

[97] Kimmel, R. and Bruckstein, A., *Shape from Shading via Level Sets*, Technion (Israel) Computer Science Dept. Report, CIS #9209 (1992).

[98] Kobbelt, L., Botsch, M., Schwanecke, U., and Seidal, H.-P., *Feature Sensitive Surface Extraction from Volume Data*, SIGGRAPH 2001, 57–66 (2001).

[99] Koepfler, G., Lopez, C., and Morel, J.-M., *A Multiscale Algorithm for Image Segmentation by Variational Method*, SIAM J. Num. Anal. 31, 282–299 (1994).

[100] Kimmel, R. and Bruckstein, A., *Shape Offsets via Level Sets*, Computer Aided Design 25, 154–162 (1993).

[101] Landau, L. and Lifshitz, E., *Fluid Mechanics*, Butterworth Heinemann (1959).

[102] LeVeque, R. and Li, Z., *The Immersed Interface Method for Elliptic Equations with Discontinuous Coefficients and Singular Sources*, SIAM J. Numer. Anal. 31, 1019–1044 (1994).

[103] Li, Z. and Lai, M.-C., *The Immersed Interface Method for the Navier–Stokes Equations with Singular Forces*, J. Comput. Phys. 171, 822–842 (2001).

[104] Lax, P. and Wendroff, B., *Systems of Conservation Laws*, Comm. Pure Appl. Math. 13, 217–237 (1960).

[105] LeVeque, R., *Numerical Methods for Conservation Laws*, Birhäuser Verlag, Boston, 1992.

[106] Liu, X.-D., Fedkiw, R., and Kang, M., *A Boundary Condition Capturing Method for Poisson's Equation on Irregular Domains*, J. Comput. Phys. 160, 151–178 (2000).

[107] Liu, X.-D., Osher, S., and Chan, T., *Weighted Essentially Non-Oscillatory Schemes*, J. Comput. Phys. 126, 202–212 (1996).

[108] Lorenson, W. and Cline, H., *Marching Cubes: A High Resolution 3D Surface Construction Algorithm*, Computer Graphics 21, 163–169 (1987).

[109] Malladi, R., Sethian, J., and Vemuri, B., *A Topology Independent Shape Modeling Scheme*, in Proc. SPIE Conf. Geom. Methods Comput. Vision II 2031, 246–258 (1993).

[110] Markstein, G., *Nonsteady Flame Propagation*, Pergamon Press, Oxford (1964).

[111] Marquina, A. and Osher, S., *Explicit Algorithms for a New Time Dependent Model Based on Level Set Motion for Nonlinear Deblurring and Noise Removal*, SIAM J. Sci. Comput. 22, 387–405 (2000).

[112] McMaster, W., *Computer Codes for Fluid-Structure Interactions*, Proc. 1984 Pressure Vessel and Piping Conference, San Antonio, TX, LLNL UCRL-89724 (1984).

[113] Menikoff, R., *Errors When Shock Waves Interact Due to Numerical Shock Width*, SIAM J. Sci. Comput. 15, 1227–1242 (1994).

[114] Merriman, B., Bence, J., and Osher, S., *Motion of Multiple Junctions: A Level Set Approach*, J. Comput. Phys. 112, 334–363 (1994).

[115] Mulder, W., Osher, S., and Sethian, J., *Computing Interface Motion in Compressible Gas Dynamics*, J. Comput. Phys. 100, 209–228 (1992).

[116] Mulpuru, S. and Wilkin, G., *Finite Difference Calculations of Unsteady Premixed Flame–Flow Interactions*, AIAA Journal 23, 103–109 (1985).

[117] Mumford, D. and Shah, J., *Optimal Approximation by Piecewise Smooth Functions and Associated Variational Problems*, Comm. Pure Appl. Math. 42, 577–685 (1989).

[118] Nguyen, D., Fedkiw, R., and Jensen, H., *Physically Based Modelling and Animation of Fire* SIGGRAPH (2002).

[119] Nguyen, D., Fedkiw, R., and Kang, M., *A Boundary Condition Capturing Method for Incompressible Flame Discontinuities*, J. Comput. Phys. 172, 71–98 (2001).

[120] Nielsen, M., Johansen, P., Olsen, O., and Weickert, J., eds., *Scale Space Theories in Computer Vision*, in Lecture Notes in Computer Science 1682, Springer-Verlag, Berlin (1999).

[121] Noh, W., *CEL: A Time-Dependent, Two Space Dimensional, Coupled Eulerian–Lagrange Code*, Methods in Comput. Phys. 3, Fundamental Methods in Hydrodynamics, 117–179, Academic Press, New York (1964).

[122] Noh, W., *Errors for Calculations of Strong Shocks Using an Artificial Viscosity and an Artificial Heat Flux*, J. Comput. Phys. 72, 78–120 (1978).

[123] Osher, S., *A Level Set Formulation for the Solution of the Dirichlet Problem for Hamilton–Jacobi Equations*, SIAM J. Math. Anal. 24, 1145–1152 (1993).

[124] Osher, S., Cheng, L.-T., Kang, M., Shim, H., and Tsai, Y.-H., *Geometric Optics in a Phase Space and Eulerian Framework*, LSS Inc. #LS-01-01 (2001), J. Comput. Phys. (in press).

[125] Osher, S. and Rudin, L., *Feature-Oriented Image Processing Using Shock Filters*, SIAM J. Num. Anal. 27, 919–940 (1990).

[126] Osher, S. and Sethian, J., *Fronts Propagating with Curvature Dependent Speed: Algorithms Based on Hamilton–Jacobi Formulations*, J. Comput. Phys. 79, 12–49 (1988).

[127] Osher, S. and Shu, C.-W., *High Order Essentially Non-Oscillatory Schemes for Hamilton–Jacobi Equations*, SIAM J. Numer. Anal. 28, 902–921 (1991).

[128] Peigl, L. and Tiller, W., *The NURBS book*, Berlin, Springer-Verlag, 2nd ed. (1996).

[129] Pember, R., Bell, J., Colella, P., Crutchfield, W., and Welcome, M., *An Adaptive Cartesian Grid Method for Unsteady Compressible Flow in Irregular Regions*, J. Comput. Phys. 120, 278–304 (1995).

[130] Peng, D., Merriman, B., Osher, S., Zhao, H.-K., and Kang, M., *A PDE-Based Fast Local Level Set Method*, J. Comput. Phys. 155, 410–438 (1999).

[131] Perona, P. and Malik, R., *Scale Space and Edge Detection Using Anisotropic Diffusion*, IEEE Trans. on Pattern Anal. Mach. Intell. 12, 629–639 (1990).

[132] Peskin, C., *Numerical Analysis of Blood Flow in the Heart*, J. Comput. Phys. 25, 220–252 (1977).

[133] Peyret, R. and Taylor, T., *Computational Methods for Fluid Flow*, Springer-Verlag, NY (1983).

[134] Puckett, E., Almgren, A., Bell, J., Marcus, D., and Rider, W., *A High Order Projection Method for Tracking Fluid Interfaces in Variable density Incompressible flows*, J. Comput. Phys. 130, 269–282 (1997).

[135] Qian, J., Tryggvason, G., and Law, C., *A Front Method for the Motion of Premixed Flames*, J. Comput. Phys. 144, 52–69 (1998).

[136] Raad, P., Chen, S., and Johnson, D., *The Introduction of Micro Cells to Treat Pressure in Free Surface Fluid Flow Problems*, J. Fluids Eng. 117, 683–690 (1995).

[137] Rogers, D., *An Introduction to NURBS*, Morgan Kaufmann (2000).

[138] Rosen, J., *The Gradient Projection Method for Nonlinear Programming, Part II. Nonlinear Constraints*, J. SIAM 9, 514–532 (1961).

[139] Rouy, E. and Tourin, A., *A Viscosity Solutions Approach to Shape-From-Shading*, SIAM J. Num. Anal. 29, 867–884 (1992).

[140] Rudin, L., *Images, Numerical Analysis of Singularities, and Shock Filters*, Ph.D. thesis, Comp. Sci. Dept., Caltech #5250:TR:87 (1987).

[141] Rudin, L. and Osher, S., *Total Variational Based Image Restoration with Free Local Constraints*, in Proc. IEEE Int. Conf. Image Proc., Austin, TX, IEEE Press, Piscataway, NJ, 31–35 (1994).

[142] Rudin, L., Osher, S., and Fatemi, E., *Nonlinear Total Variation Based Noise Removal Algorithms*, Physica D 60, 259–268 (1992).

[143] Russo, G. and Smereka, P., *A Remark on Computing Distance Functions*, J. Comput. Phys. 163, 51–67 (2000).

[144] Ruuth, S., Merriman, B., and Osher, S., *A Fixed Grid Method for Capturing the Motion of Self-Intersecting Interfaces and Related PDEs* J. Comput. Phys. 163, 1–21 (2000).

[145] Sapiro, G., *Geometric Partial Differential Equations and Image Analysis*, Cambridge U. Press (2001).

[146] Sethian, J., *An Analysis of Flame Propagation*, Ph.D. Thesis, University of California at Berkeley (1982).

[147] Sethian, J., *Curvature and the Evolution of Fronts*, Comm. in Math. Phys. 101, 487–499 (1985).

[148] Sethian, J., *A Fast Marching Level Set Method for Monotonically Advancing Fronts*, Proc. Nat. Acad. Sci. 93, 1591–1595 (1996).

[149] Sethian, J., *Fast Marching Methods*, SIAM Review 41, 199–235 (1999).

[150] Shu, C.W. and Osher, S., *Efficient Implementation of Essentially Non-Oscillatory Shock Capturing Schemes*, J. Comput. Phys. 77, 439–471 (1988).

[151] Shu, C.W. and Osher, S., *Efficient Implementation of Essentially Non-Oscillatory Shock Capturing Schemes II (two)*, J. Comput. Phys. 83, 32–78 (1989).

[152] Smereka, P., *Spiral Crystal Growth*, Physcia D 138, 282–301 (2000).

[153] Son, G. and Dir, V.K. *Numerical Simulation of Film Boiling near Critical Pressures with a Level Set Method*, J. Heat Transfer 120, 183–192 (1998).

[154] Spiteri, R. and Ruuth, S., *A New Class of Optimal High-Order Strong-Stability-Preserving Time Discretization Methods*, SIAM J. Numer. Anal. (in press).

[155] Staniforth, A. and Cote, J., *Semi-Lagrangian Integration Schemes for Atmospheric Models—A Review*, Monthly Weather Review 119, 2206–2223 (1991).

[156] Steinhoff, J., Fang, M., and Wang, L., *A New Eulerian Method for the Computation of Propagating Short Acoustic and Electromagnetic Pulses*, J. Comput. Phys. 157, 683–706 (2000).

[157] Strikwerda, J., *Finite Difference Schemes and Partial Differential Equations*, Wadsworth & Brooks/Cole Advanced Books and Software, Pacific Grove, California (1989).

[158] Sussman, M. and Fatemi, E., *An Efficient Interface-Preserving Level Set Redistancing Algorithm and Its Application to Interfacial Incompressible Fluid Flow*, SIAM J. Sci. Comput. 20, 1165–1191 (1999).

[159] Sussman, M., Fatemi, E., Smereka, P., and Osher, S., *An Improved Level Set Method for Incompressible Two-Phase Flows*, Computers and Fluids 27, 663–680 (1998).

[160] Sussman, M., Smereka, P., and Osher, S., *A Level Set Approach for Computing Solutions to Incompressible Two-Phase Flow*, J. Comput. Phys. 114, 146–159 (1994).

[161] Steinhoff, J. and Underhill, D., *Modification of the Euler Equations for "Vorticity Confinement": Application to the Computation of Interacting Vortex Rings*, Physics of Fluids 6, 2738–2744 (1994).

[162] Tang, B., Sapiro, G., and Caselles, V., *Color Image Enhancement via Chromaticity Diffusion*, IEEE Transactions on Image Processing 10, 701–707 (2001).

[163] Teng, Z.-H., Chorin, A., and Liu, T.-P., *Riemann Problems for Reacting Gas with Applications to Transition*, SIAM J. Appl. Math 42, 964–981 (1982).

[164] Toro, E., *Riemann Solvers and Numerical Methods for Fluid Dynamics*, Springer-Verlag (1997).

[165] Tryggvason, G, Bunner, B., Juric, D., Tauber, W., Nas, S., Han, J., Al-Rawahi, N., and Jan, Y.-J., *A Front Tracking Method for the Computations of Multiphase Flow*, J. Comput. Phys. V. 169, 708–759 (2001).

[166] Tsitsiklis, J., *Efficient Algorithms for Globally Optimal Trajectories*, Proceedings of the 33rd Conference on Decision and Control, Lake Buena Vista, LF, 1368–1373 (December 1994).

[167] Tsitsiklis, J., *Efficient Algorithms for Globally Optimal Trajectories*, IEEE Transactions on Automatic Control 40, 1528–1538 (1995).

[168] Unverdi, S.O. and Tryggvason, G., *A Front-Tracking Method for Viscous, Incompressible, Multi-Fluid Flows*, J. Comput. Phys. 100, 25–37 (1992).

[169] Van Dyke, M., *An Album of Fluid Motion*, The Parabolic Press, Stanford (1982).

[170] Vese, L. and Osher, S., *Numerical Methods for p-Harmonic Flows and Applications to Image Processing*, UCLA CAM Report 01-22, SIAM J. Num Anal. (in press).

[171] Welch, S. and Wilson, J., *A Volume of Fluid Based Method for Fluid Flows with Phase Change*, J. Comput. Phys. 160, 662–682 (2000).

[172] Whitaker, R., *A Level Set Approach to 3D Reconstruction from Range Data*, Int. J. of Comput. Vision (1997).

[173] Williams, F.A., *The Mathematics of Combustion*, (J.D. Buckmaster, ed.), SIAM, Philadelphia, PA, 97–131 (1985).

[174] Zhang, Y.L., Yeo, K.S., Khoo, B.C., and Wang, C., *3D Jet Impact of Toroidal Bubbles*, J. Comput. Phys. 166, 336–360 (2001).

[175] Zhao, H.-K., Chan, T., Merriman, B., and Osher, S., *A Variational Level Set Approach to Multiphase Motion*, J. Comput. Phys. 127, 179–195 (1996).

[176] Zhao, H.-K., Osher, S., and Fedkiw, R., *Fast Surface Reconstruction Using the Level Set Method*, 1st IEEE Workshop on Variational and Level Set Methods, 8th ICCV, Vancouver, 194–202 (2001).

[177] Zhao, H.-K., Osher, S., Merriman B., and Kang, M., *Implicit and Nonparametric Shape Reconstruction from Unorganized Data Using a Variational Level Set Method*, Comput. Vision and Image Understanding 80, 295–314 (2000).

Index

Applied Mathematical Sciences

(continued from page ii)

(continued on next page)

Applied Mathematical Sciences

(continued from previous page)